Relevance of Local Environmental Knowledge in the Era of Globalization

Relevance of Local Environmental Knowledge in the Era of Globalization

K. Anil Kumar

SERIALS PUBLICATIONS
NEW DELHI (INDIA)

First Published - 2011

ISBN: 978-81-8387-395-6

Published by

SERIALS PUBLICATIONS
4830/24, Prahlad Street, Ansari Road
Darya Ganj, New Delhi-110002 (India)
Phone : 23245225 Fax : 91-11-23272135
E-mail: serials@sify.com

Dedicated to

The Konda Reddis' of Utalapalem Village, East Godavari District

Contents

Preface

Generally, when we talk or think about science, we refer to western or industrial society. In the context of scientific development, we prefer to discuss more about the so-called modern scientific innovations, assuming that science is only there in those societies and implying that scientific rigor or interest is absent in other societies. But we also need to look at indigenous science and technology, particularly when an existing body of knowledge is available; perhaps it is better to develop on it instead of replacing it in the name of scientific development. Again, in the present day context of globalization process and patent acts, there is a need to preserve and protect indigenous wisdom, knowledge and science. Although indigenous peoples constitute one of the largest vulnerable segments in contemporary society, they and their knowledge systems have been marginalized mainly because of the craze for modernity and globalization. The distinct culture of indigenous people, their identity, economic activities, religious beliefs, notions, and traditional ways of managing natural resources are often regarded as backward and superstition. They are considered to be absolutely incompatible with modern society and development. But, indigenous communities played an important role in generating knowledge based on the understanding of their environment, devising mechanisms to conserve and sustain their natural resources and establishing community-based organisation that serve as a forum for identifying problems and dealing with them through local-level experimentation, innovation and exchange of information with other societies (Warren, 1992).

For centuries, indigenous communities have developed ingenious methods of utilization of natural resources for their survival. The development of indigenous knowledge systems including Management of natural environment has been a matter of survival for the people who generated these systems. The subsistence systems depends usually upon nature of habitat, accessibility of natural resources degree of attainment of knowledge and technology for utilization of available resources.

Indigenous communities more or less have been passing through various stages of economic evolution/development, starting with hunting and gathering to settled cultivation and market economy.

With the rapid environmental, social, economic and political changes occurring in many areas inhabited by indigenous people comes the danger that the Indigenous Knowledge (IK) they possess will be overwhelmed and lost forever. Younger generations are acquiring different values and lifestyles as a result of exposure to global and national influences, and traditional communication networks are breaking down, meaning that Elders are dying without passing their knowledge onto their children. In some cases, the actual existence of indigenous people themselves is threatened. Indigenous knowledge forms an important part of the heritage. It is imperent to recognize the value of indigenous/local which supports the survival of cultural and biological diversity. Efforts should be made to capture and preserve it so that it can be documented and shared.

Against this backdrop, the present book, which is based on my Ph.D thesis, made an attempt to record Konda Reddis' Indigenous Environmental Knowledge which is very much important for thesis sustenance. Among the Konda Reddis of East Godavari District, forest knowledge is found to be very useful in identifying and locating resources and that sustainable practices ensured continuity of there resources. The book highlights the community's indigenous agricultural knowledge and the changes over time. The study also identifies spiritual knowledge and beliefs as fundamental for developing indigenous worldviews and environmental attitudes and values, which play a very important role in their everyday activities. The study concludes by saying that Indigenous Knowledge (IK) is very useful for self sustenance of the community, in order to reduce hunger and poverty among indigenous people.

The present study would not have been possible without the support that I received from different sources. First and Foremost, I would like to express my deep sense of gratitude and heartfelt thanks to my honourable supervisor, Prof. K. K. Misra, for his guidance, invaluable suggestions, encouragement, constant support and patience throughout my research work, which I can never forget. His Knowledge and intellectual vision have widened the scope of my thinking. I owe a deep sense of gratitude to him for being part of my supervisor during a part of my M.Phil and Ph.D Thesis, with a sense of pride and privilege.

I would like to thank respected teacher Prof. P. Venkata Rao, for his kind words of encouragement and the kindly support. I wish to acknowledge with gratitude the timely support rendered by Dr. Siva Prasad and Dr. Laxminarayana of my Ph.D Advisory Committee Members. I thank them for their encouragement and support. This is also an occasion to thank respected teachers Prof. Sudhakar Rao, Dr. B.V Sharma, Dr. G. Tarakan and Md. Abdul Munnaf for their unstinted support, cooperation and warmth. My sincere thanks are due to Prof. Subhadra Channa (University of Delhi, Delhi) and Prof. V. Gangadharam (Sri Ventateswara University, Tirupati) who were externals of my thesis. I thank them for thesis detailed Comments which gave me further scope to improve my present work.

I would also like to acknowledge with much appreciation the crucial role of the Konda Reddi community of Utalapalem Village who spent their precious time despite their tight schedules and served to be the major contributors to make this research successful. I am indebted for the hospitality and many a kindness shown to me during the course of my fieldwork. In this regard, I fail my duty if I do not acknowledge the timely help that I had received from Radha Krishna, Satibabu, Venkata Reddi, Bodhi Reddi, Devalamma, and Nagesh. Therefore, with great pleasure I dedicate this book to Konda Reddi community of Utalapalem Village.

My special thanks to former head of the department Prof. D.V. RaghavaRao for his kind words and immense support every time.

My special thanks to Dr. Sowmya, Nandu, Usha, for their needful help.

I extend my heart felt thanks to the Library staff (Indira Gandhi Memorial Library UOH), NIRD library, TCRTI and Osmania library.

I am thankful to ICSSR for providing a Doctotal fellowship, which enabled me to complete the research work in time.

My close friends Dr. Romesh Singh, B.K. Srinivas, Dr. Vamsi, Dr. Shekar and Raj have always been a constant source of inspiration and encouragement all through my stay in the University. Iam very thankful to all my seniors, juniors, classmates and friends for their timely and invaluable help.

I remain indebted to my parents and family members for a never-ending support.

Iam forever greatful to my father-in-law and Mother-in-law who gently offer unconditional support at each turn of the road; and to my wife Parvathi and children Ridhi and Ribhu for being source of inspiration and their concern to complete this book.

Last but not least, I extended my heartfelt thanks to Serials Publications, New Delhi for publishing my work promptly in stipulated period.

Dr. K. Anil Kumar

1

Introduction

Human dependence on the environment around made them acquire the knowledge and value of many plants, animals and other natural resources by trial and error. Consequently, humans became the storehouse of knowledge of many useful as well as harmful plants and animals, accumulated and enriched through generations. This knowledge was passed on from one generation to another, usually verbally, without any written document is called indigenous knowledge or local environmental knowledge. Indigenous knowledge or local environmental knowledge is the knowledge that people have gained through inheritance from their ancestors. Over centuries, indigenous people around the world have developed their own, locality-specific knowledge and practice, which is an important part of the lives of the people. It is a people-derived science, and it represents people's creativity, innovations and skills.

The history of local environmental knowledge is as old as the human race. This knowledge has always been very important for the local people who generate it; primarily, it is a matter of survival for them. Local knowledge that is unique to a culture or society. It is also known as indigenous knowledge folk knowledge, people's knowledge, traditional wisdom or traditional science. Moreover, local knowledge is also a key element of the social capital of the poor and constitutes their main asset in their efforts to gain control of their own lives. Indigenous Knowledge Systems (IKS) are an important part of the lives of the poor. It is the basis for decision-making of communities in food security, human and animal health, education, and natural resource management. Indigenous Knowledge Systems point to how indigenous people manipulate their knowledge, which has accumulated, evolved and been practised for generations. They epitomise the relationship and interaction between indigenous peoples and their natural surroundings. In due course of time, many experts like (Brokensha *et al.*, 1980: Warren *et al.*, 1995: Chambers, 1983) etc. have highlighted the valuable elements of local knowledge that need to be nurtured and strengthened for sustainable development and then employed for the welfare of the wider population. The potential contribution of indigenous knowledge to locally manage, sustainable and cost-effective survival strategies should be promoted in the development process.

After independence over five decades of mainstream development, have failed to recognize the existing "knowledge-base" of the communities. Planners have designed development interventions that are entirely divorced from the local reality, consequently. In this manner a vast chunk of the said local environmental knowledge is rapidly getting depleted, as local knowledge systems are primarily based on oral traditions. The recent international interest in indigenous knowledge (IK) as an important inputs to on going planning process for sustainable development and environmental conservation means that a great deal of field research has to be carried out on these indigenous communities, which until recently, was confined to a few individuals and institutions, with unsatisfactory results therefore, a growing number of scientists and development professionals are now gradually recognizing that indigenous knowledge provides a valuable resource for sustainable development. As such today, importance and contributions of indigenous knowledge systems and practices has invited special attention from various disciplines such as biology, medicine (human and animal), pharmacology, agronomy, soil sciences and various social sciences. In recent development theory and practice, many researchers all over the world are now striving to explore, know, record and use what the indigenous people already know about using and conserving biological resources. Indigenous knowledge systems (IKS) are now being examined by academicians, development planners, and contributors to provide alternative development approaches (Warren *et al.*, 1989).

Defining Indigenous Knowledge

Knowledge is processed and summarized information on any subject with definite inter/extrapolated inferences. Knowledge is an appreciation of the possession of interconnected details, which in isolation, are of lesser value. Knowledge refers to what one knows and understands. Knowledge is often categorized as unstructured, structured, explicit or tacit. However, what is meant by 'indigenous knowledge' is not very clear at the moment when it is rapidly coming into current use with development planners. A variety of terms have been used in development literature to refer to the collective knowledge of local people. Indigenous knowledge has often been treated as a synonym to local or traditional knowledge; and this knowledge has variously been referred to various researchers as ethnoscience, people's science, village science (Barker, 1977: Chambers, 1983), rural people's knowledge (Chambers, 1983), indigenous technical knowledge (ITK) (IDS 1979), local knowledge (Korten and Uphoff, 1981, quoted in Chambers, 1983), indigenous knowledge or local knowledge (Dei *et al.*, 2000: Semali and Kincheloe, 1999), Indigenous/local knowledge (Flora, 1992: Kloppenburg, 1991: Warren *et al.*, 1995). Ethno-ecology, folk-ecology, traditional environmental knowledge (TEK), traditional ecological knowledge, indigenous (ecological) knowledge, knowledge of the land, customary law (Johnson,

1992), folk science, people's knowledge, traditional knowledge, citizen science (in Europe) (Sillitoe, 2002b), indigenous environmental knowledge (IEK), traditional knowledge systems (TKS), cunning intelligence (Ellen, 2002), endogenous knowledge, autochthonous knowledge, sustainable knowledge (Antweiler, 2004), etc., all terms for knowledge belonging to various groups of ethnic grassroots people. Nonetheless, they all share a certain common idea and address the same broad issues, which are used interchangeably to refer to that knowledge which is generated and transmitted by local communities, over time, in an effort to cope with their own agro-ecological and socio-economic environments.

Indigenous knowledge has recently become popular with many social scientists who point out that indigenous peoples have their own effective 'science' and it is necessary to understand their knowledge and management systems (Kloppenburg, 1991: DeWalt, 1994: Sillitoe, 2002). The diverse terms as stated above reflect several and often implicit epistemological assumptions and diverse socio-political agendas. Each of the terms has its drawbacks (Antweiler, 2004). For instance, people's science is used to describe the knowledge system of a group of rural people. But in addition, it has been used to describe not the science of the people, but science for the people. The term ethnoscience is used to describe indigenous knowledge of a self- contained community and also refers to the methodology for eliciting, translating and interpreting the knowledge system of that particular culture (Brokensha *et al.*, 1980: Chambers, 1983). Some are critical about the use of the word indigenous on the ground that it is difficult to determine the status of indigene and suggest other terms such as traditional or local etc when referring to this knowledge (Sillitoe and Bicker, 2004). But there are objections to the use of these terms also, as tradition suggests timeless continuity and the use of the term traditional raises questions regarding the cultural dynamics of such knowledge systems, as knowledge is ever changing and dynamic (Johnson, 1992: Sillitoe, 2002). Thus, there is an ongoing and often tedious terminological debate. Each term has been generated in a particular context and none of them is universally accepted. All have their uses, advantages and disadvantages, noting marginal differences in connotations (although many with some degree of semantic overlap). Ellen (2002) opines that this debate is ultimately irresolvable and therefore arguably futile.

Even though there is no universally accepted definition of indigenous knowledge in literature, there are a number of definitions of "local knowledge", and most of them are confronted by some fundamental problems of differentiating it from other types of knowledge. This is in part due to the differences in background and perceptions of the various authors who come from varied fields ranging from social anthropology to agricultural engineering. Nevertheless, the various definitions also have some common traits; following which can be defined as "a cumulative body of knowledge and beliefs originating from and naturally produced in an area" (Chambers, 1983) or as "a cumulative body of knowledge generated and evolved over time, representing generations of creative thought and actions within individual societies in an ecosystem of continuous residence with an effort of coping with the ever changing agro-ecological and socio-economic environment (Fernandez, 1994: LaDuke, 1994: Laws and Luning, 1996: Serrano, 1996: Warren, 1996: Darkoh, 1996 : from Mbaiwa ,2002). For Grenier (1998) it is "the unique, traditional, local knowledge existing within and developed around specific conditions of women and men indigenous to a particular geographic area".

Indigenous knowledge can also refer to the 'technical' insight or wisdom gained and developed by people in a particular locality through years of careful observation and experimentation with the natural phenomena around them. Indigenous knowledge refers to the knowledge of indigenous peoples as well as other well defined communities. Furthermore as Maurial (1999) states "Indigenous knowledge is *local* because it is the result of the quotidian interactions in indigenous people's territories".

Flora (1992) draws the relation of indigenous knowledge to the development of technologies. "Part of indigenous knowledge consists of technologies developed over decades of adjusting farming systems to local agro-climatic and social conditions. And in some circumstances, local knowledge also consists of knowing how to keep conditions of productivity over the long run, rather than maximizing productivity in years of optimal conditions". "Indigenous knowledge reflects the dynamic way in which the residents of an area have come to understand themselves in relationship to their natural environment and how they organize that folk knowledge of flora and fauna, cultural beliefs, and history to enhance the quality of their lives" (Semali and Kincheloe's, 1999). Warren *et al.* (1995) explore the connection between indigenous knowledge and cultural and sociological background, and say that indigenous knowledge or local knowledge is the one that "is unique to a given culture or society.

All of the above said concepts of indigenous knowledge systems delineate a cognitive structure in which theories and perceptions of nature and culture are conceptualized. Thus, it includes definitions, classification and concepts of the physical, natural, social, economic and ideational environments. The dynamics of indigenous knowledge systems takes place on two different levels, the cognitive and the empirical. On the empirical level, indigenous knowledge systems are visible in institutions, artifacts and technologies. Indigenous technical knowledge is of a practical nature, concerned with operationalized local thinking in such fields as agriculture, fisheries, health, horticulture, and forestry (Brouwer, 1998).

According to the literature adopted from Ellen and Harris (1996) IK is:

Local, in that it is rooted in a particular community and situated within broader cultural traditions; it is a set of experiences generated by people living in those communities. Separating the technical from the non-technical, the rational from the non-rational could be problematic. Therefore, when transferred to other places, there is a potential risk of dislocating indigenous knowledge. Generalized characteristic of indigenous knowledge can be summarized as below :

- *Tacit* knowledge and, therefore, not easily codifiable.
- *Transmitted orally,* or through imitation and demonstration. Codifying it may lead to the loss of some of its properties.
- *Experiential rather than theoretical knowledge.* Experience and trial and error, tested in the rigorous laboratory of survival of local communities constantly reinforce IK.
- *Learned through repetition,* which is a defining characteristic of tradition even when new knowledge, is added. Repetition aids in the retention and reinforcement of IK.
- *Constantly changing,* being produced as well as reproduced, discovered as well as lost; though it is often perceived by external observers as being somewhat static.

Consolidation of Indigenous Knowledge Studies

Awareness of indigenous/local knowledge (IK/ LK) has been steadily growing in the academic world, in social sciences and natural sciences. "A growing number of scientists and policy makers are aware today of the contribution indigenous knowledge (IK) can make to a more sustainable development" (Viergever, 1999). IK also seems to be relevant to the scientific world for a number of reasons including issues of protection of biodiversity (Iwanaga, 1998), protection of Intellectual Property Rights (IPR) of the rural communities (RAFI, 2000: RAFI/ UNDP, 1995), and the fact that IK could be used as the starting point in the construction of a truly alternative agriculture (Flora, 1992: Kloppenburg, 1991).

Consequently many research and development institutions devoted to agricultural development look at indigenous knowledge as a key component of sustainable agricultural practices; and are therefore researching and cataloguing existing indigenous knowledge. The Center for Indigenous

Knowledge for Agriculture and Rural Development (CIKARD), established in 1987 at Iowa State University, is an example. CIKARD "focuses its activities on documenting and preserving the indigenous knowledge of farmers and other rural people around the globe" (Warren and McKiernan 1995). Another institute, viz, Consultative Group on International Agricultural Research Centers (CGIAR), began documentation of local/indigenous knowledge for enhancing technology. For e.g., technologists from the International Potato Center (CIP) in Peru worked with local farmers to develop storage technology for preserving potato seeds (Fujisaka, 1995).

Similarly, various international bodies, such as the International Union for the Conservation of Nature and Natural Resources, the International Biodiversity Convention, the Agenda 21 Report of the UN Conference on Environment and Development have recognized the importance of indigenous knowledge (Davis 1993). The World Commission on Environment and Development regards indigenous knowledge as a significant variable and a critical factor in achieving sustainable development. It is now widely acknowledged that traditional communities "are the repositories of vast accumulations of traditional knowledge and experience...Their disappearance is a loss for the larger society, which would learn a great deal from their traditional skills in sustainably managing very complex ecological systems" (WCED, 1987).

Indigenous versus Local Knowledge

Indigenous people are often defined as the original inhabitants of a particular geographic location, who have a culture and belief system distinct from the international system of knowledge (e.g., the Tribal, Native, First, or Aboriginal people of an area). Some, however, feel that such a definition is too narrow, as it excludes people who may have lived in an area for a long period of time but are not the original inhabitants. This has led to widespread use of the term local knowledge which refers to the knowledge possessed by any group living in a particular area for a long period of time. In this approach, it is not necessary to know if the people in question are the original inhabitants of an area, the important thing here is to note how the local people—aboriginal or non-aboriginal—in a particular area view and interact with their environment. In this book, for simplicity the two terms are taken as synonymous and used as interchangeable words. It is in this regard that a wider definition of local knowledge, has been made that includes cultural and technical knowledge. It is also 'rural peoples' knowledge or 'local technical knowledge' that links traditional technical knowledge with the social and political knowledge and skills of the local people.

A Wider Perspective on Local Knowledge

Traditional knowledge is a cumulative body of knowledge, know-how, practices and representations maintained and developed by peoples with extended histories of interaction with the natural environment. These sets of understandings, interpretations and meanings are part and parcel of a cultural complex that encompasses language, naming and classification systems, resource use practices, ritual, spirituality and worldview (Berlin, 1992: Balick, M and P Cox, 1997). Traditional knowledge provides the basis for local-level decision-making about many fundamental aspects of day-to-day life: hunting, fishing, gathering, agriculture and husbandry; preparation, conservation and distribution of food; location, collection and storage of water; struggles against disease and injury; interpretation of meteorological and climatic phenomena; confection of clothing and tools; construction and maintenance of shelter; orientation and navigation on land and sea; management of ecological relations of society and nature; adaptation to environmental/social change; and so on and so forth.

It is obvious that local or indigenous knowledge, like any other form of knowledge has been developed within specific cultural groups over a specific period of time and within specific environmental and social settings. At the same time, history has demonstrated how knowledge has been actively shared and exchanged among societies, and in this matter, holders of traditional knowledge do not differ. In this respect modern science is also not different. It is also anchored in a specific worldview and, more to the point, a specific view about man's relation to nature that is strongly instrumental (Keith Thomas, 1983). In contrast, the worldview embraced by indigenous knowledge holders typically emphasizes the symbiotic nature of the relationship between humans and the natural world. Rather than opposing man and nature as in occidental thought, traditional knowledge holders tend to view people, animals, plants and other elements of the universe as interconnected by a network of social relations and obligations (Ann Fienup-Riordan, 1990).

Holistic cosmologies that intertwine elements that are ecological and social, as well as empirical and spiritual, have sometimes confounded scientists who may seek to separate 'fact' from 'superstition'. Such a dualistic approach, however, contains certain dangers. Practices that appear in the first instance as superstitious to the outside observer, may be an additional knowledge about the environment and culture which may provide appropriate and empirically sound ways of coping with environmental problems. Furthermore, practices may have latent meanings that may only be revealed through a better understanding of the culture as a whole. In general, by isolating elements from a worldview that interweaves empirical, spiritual, social and other components, as indigenous knowledge does, one tends to misrepresent both the elements and the whole.

While it is possible to indicate a loose set of properties of traditional knowledge, such characteristics must be viewed as provisional rather than definitive and mutually exclusive. Traditional knowledge is a broad umbrella, encompassing configurations as diverse as decision-making strategies among shepherds (Agrawal, 1993), multiple tree cropping systems of small-holders (Thrupp, 1989), techniques for domestication of crops (Reed 1977; Rhoades 1989), and plant classification systems (Brush, 1980). "Ethnoscience" is a scientific approach to traditional knowledge, based on the work of Harold Conklin among the Hanunoo of the Philippines in the 1950, where by study of responses to plants, diseases, soils, animals, and agriculture, scientists developed an appreciation of the coherence of indigenous knowledge systems, and their attunement to local environment. Ethnobotanists discovered cases in which the number of plant species recognized by local communities was greater than the number of scientific species recognized in an area. Such realization has not been limited to academic scientists, but extended to pharmaceutical and agricultural companies which led to concerns about bioprospecting and indigenous property rights.

As ethnobotany grew, scientific interest increased not only among those working in the field of anthropology but also among those in the field of biological systematics, structural linguistics, cognitive psychology, and logic. Simultaneously, debates continued on the extent of universality of their classification systems (Berlin, 1992: Ellen, 1998: Atran, 1991).

In the past, most of this work was academic in nature, examining traditional knowledge for its own sake, but from the 1980's this became an applied science & researchers in multilateral and bilateral development agencies began to recognize the significance of indigenous knowledge for sustainable development, both for environmental conservation and technologies for agricultural productivity. For example, scientists in the international CGIAR (Consultative Group on International Agricultural Research) began to value participatory technology development, using the indigenous knowledge of local populations as a starting point. As ecological concerns gained currency these approaches were extended to the management of natural resources, conservation strategies, and interdisciplinary collaborations that relied heavily on local knowledge. In practical terms, scientists began to work closely with indigenous communities to promote agriculture and ecological practices. Such efforts are likely to increase in importance in future because of the recognition that many environmental problems are mostly local in nature.

These relationships between scientists and indigenous communities have been criticized by some enjoying hegemonic power relationships embedded in certain forms of knowledge.

Western science, insofar as it embodies cultural constructions that subordinate traditional populations and tribal groups, is viewed as an enemy of indigenous knowledges as these promote political aims of empowerment and land rights for these groups. Although they value indigenous knowledge for its own sake, their ideas of the land rights of local communities and environmental conservation has become contentious as it is based on a notion of an "ecologically noble savage" rather than consistent findings on the relationship between the use of indigenous knowledge and ecological well-being (Zent, 2000).

In recent years, it is increasingly acknowledged that traditional knowledge (TK), through modern ethnobotanical research, is helping scientists to understand management of biodiversity and natural forest management by crop domestication and breeding. It is realized that traditional knowledge also leads to new appreciation of the principles and practices of agroecology, agroforestry, crop rotations, pest and soil management, and other areas of agricultural science.

Ethnobiology is a branch of ethnoscience devoted to the study of the reciprocal interactions between people and biological organisms as observed by the local communities. Traditional knowledge profoundly influences ethnobiology and ethnoscience which are now often adopted by science and re-applied in contemporary contexts and through contemporary management (Cunningham, 2001). Traditional knowledge has also helped modern science in many other areas, most notably in taxonomy, medicine, agriculture, natural resource management, and conservation. For example, in taxonomy many species new to science have been pointed out by traditional peoples knowledgeable in the flora and fauna of their environment *e.g.*, new species of primates were recently discovered in Central and South America, new ungulates in Southeast Asia, and new plant species throughout the tropics under the guidance of traditional people and their knowledge.

Similarly, medicine is influenced by traditional knowledge in many ways and today it is estimated that as many as 80% of the world's people depend on traditional medicine for their primary health care needs. The combination of traditional and scientific knowledge is evidenced *e.g.*, in the USA where in 1985 25% of all prescriptions contained plant materials (Farnsworth and Soejarto, 1985). The use of traditional knowledge in bioprospecting for new pharmaceuticals is an active scientific pursuit (Chadwick and Marsh, 1994). Similarly, our appreciation of the subtle and often unarticulated indigenous strategies in natural resource management has been fostered through ethnobotanical studies of indigenous knowledge. This has inestimably promoted scientific advancement in natural resource management.

Application of the ethnosciences to conservation (Cunningham, 2001) enables effective management and partnerships without which conservation is doomed. From the harvesting of individual resource to the management of entire ecosystems, lessons from local people enables the conservationists to integrate their programs with real human needs and practices. Applying ethnosciences and traditional knowledge to sustainable development programmes is a further step with great potential (Bennett, 1992) but it is also associated with problems including that of land tenure, genetic resource ownership, intellectual property rights and benefit sharing (ten Kate and Laird, 1999). Recognizing the worth and value of traditional knowledge in science also involves various issues of professional ethics (Cunningham, 1996).

From the mid -1990s onwards, there has been an emerging trend of doubting the utility of indigenous knowledge. Paul Richards, a development anthropologist working on agriculture in Africa, was sharply critical of the idea that indigenous knowledges were *systems*, coherent and static wholes that a farmer might simply apply like a technology package. He argued that a farmer's skills consisted of a set of capacities that the farmer might draw on selectively in response to changing situations, and that "good farming" consisted of flexible and creative selection.

In another contemporary work "Dismantling the Divide between Scientific and Indigenous Knowledge" (Arun Agrawal 1995) directly challenged the epistemological basis of drawing distinction between the knowledge of a Western scientist and an indigenous farmer. Since both are non-unitary, contingent on situational practices, and capable of generating generalizable theory only through an act of social will, the line becomes quite blurry. This is where the anthropology of indigenous knowledge comes closer to the sociology of scientific knowledge. Agrawal is particularly troubled by the limited databases for traditional knowledge and other projects taken up for preserving, promoting, or translating indigenous knowledge. He asks if these knowledge regimes can only be recognized as such by their scope, their scale, and local specificity. What happens when NGOs or government bodies try to 'scale them up'.?

Ethnobiologists, Roy Ellen and other contributors to a volume called *Indigenous Ecological Knowledge and its Transformations* (2000) also point, to a growing consensus that this so called indigenous knowledge is not really the *sui generis* category that we once thought. They also raise concerns about the political ambiguity of indigenous knowledge as a discourse.

Local knowledge is considerably more complex. It may include causal knowledge and knowledge of complex systems, for instance, cropping systems (about ways of farming crops), or treating environmentally induced diseases. It may relate to relationships in everyday life (model of whole society, models of 'honour', of 'marriage', of 'justice', etc.) or information of relationships within the cosmos (cosmology), or the aetiology of diseases (therapies), the creation of humankind or the origin of a community. Such complex knowledge buildup may be termed 'belief systems' and not knowledge systems. The word 'system' here indicates that these constructs are not simply aggregates of isolated concepts, but comprise interlinked concepts and their constituent elements. Within such systems, cosmologies and prototypical concepts play a major role. These are termed 'cultural models' (Antweiler, 2004).

The diversity, dynamics and patterns of indigenous knowledge need to be understood as part of the cultural system, which has been structured according to gender, age, occupation, caste, class, etc., (Sillitoe, 2002b). This is not unique to natural science. They are often implicit in works of ethnobotany, ethnozoology, human ecology and environmental anthropology. Anthropologists and other social and natural scientists who have written about 'indigenous knowledge' often make reference to the notion of 'value'. Where the concept of making local knowledge available and applicable is geared to local self-reliance, decentralization of decision-making and fair access to natural resources. This approach involves the local population who determine their own objectives, implement the measures themselves on the basis of their own view of and approaches to solving their own problems, thereafter evaluating the results themselves.

Scientific Knowledge and Local Indigenous Knowledge: Commonalities and Differences

The word local connotes the dimension of situatedness in local culture and environment (Antweiler, 2004) and the word "indigenous" implies "originating from and naturally produced in an area" (Chambers, 1983). Local-indigenous knowledge, the term used here, refers to indigenous knowledge of the local people/local knowledge of indigenous people. Here the facts, skills and capabilities pertaining to knowledge are local to the extent that they are acquired and applied by people with respect to local objectives, situations, and problems; the solutions are drawn on locally available raw materials and energy sources.

Antweiler (2004) opined that local knowledge could not be grasped by simple dichotomies such as science vs. wisdom/ belief, ethnoscience vs. technoscience or scientists vs. folk. Local knowledge may share some features with science and some with other ways forms of knowing. We must, therefore, think of it as resulting from a universal human capability, but having specific cultural and environmental situated content in every instance. According to Posey (2002, 2004) indigenous knowledge is not local knowledge, but rather an expression of universal knowledge expressed through the local. Thus, local knowledge has both its universal capability and specific situational character, as it is a locally situated form of knowledge and performance found in all societies.

There are both similarities and differences between localized traditional science and modern science, which was primarily developed in the western hemisphere. Bronowski considers the practice of science (including magic) as a fundamental characteristic of human societies: "...to me the most interesting thing about man is that he is an animal who practices art and science and, in every known society, practices both together" (Bronowski, 1978). Thus, one can probably say that both modern science and indigenous knowledge (and art) are the result of the same general intellectual process of creating order out of disorder.

There are also major differences, however, between the two kinds of science, some of them substantive and perceptual. Johannes (1989) observes that "the attitudes of many biological scientists and natural resource managers to traditional knowledge has frequently been dismissive." Accomplishments of traditional societies in such fields as agriculture cannot be denied; most domesticated species predate modern science. Nevertheless, the existence of curiosity-driven inquiry among traditional peoples has been questioned by those who regard the knowledge of other cultures as irrational, thus playing down the validity of TEK.

Fikret Berkes (1993) suggested 9 criteria in general that Traditional Ecological Knowledge differs from scientific ecological knowledge in a number of substantive ways: (1) TEK is mainly qualitative (as opposed to quantitative). (2). TEK has an intuitive component (as opposed to being purely rational). (3). TEK is holistic (as opposed to reductionist). (4). In TEK, mind and matter are considered together (as opposed to a separation of mind and matter). (5). TEK is moral (as opposed to supposedly value-free). (6). TEK is spiritual (as opposed to mechanistic). (7). TEK is based on empirical observations and accumulation of facts by trial-and-error (as opposed to experimentation and systematic, deliberate accumulation of fact). (8). TEK is based on data generated by resource users themselves (as opposed to that by a specialized cadre of researchers). (9). TEK is based on diachronic data, i.e., long time-series on information on one locality (as opposed to synchronic data, i.e., short time-series over a large area).

Does local knowledge ultimately match the knowledge gained through modern science, or is it structured entirely differently? This is an age-old topic of debate in anthropology; the debate concerned with rationality *vis a vis* various *modes of thought* (Pelissier, 1991: Fikentscher, 1995: Olson and Torrance, 1996). The corresponding practical question is whether local knowledge can be utilized within the framework of scientifically-based society and whether local knowledge represents the holistic counter model to science and as such can be used as a tool to criticize practical measures founded on analytical modern science.

It is remarkable that most of the characterizations of local knowledge as given in the above criteria are negative while being defined in contrast to scientific knowledge (Mc Corkle, 1989). But that is natural as local and scientific knowledge are neither completely different nor entirely the same so they display both commonalities and differences at different points.

One fundamental difference between these two ways of knowing is that the sciences seek information, which is transferable to any spatial or social situation (*immutable mobiles*), i.e. which is not context-bound. As a result, scientists know a great deal about small sections of reality. By contrast, local knowledge systems seek specific knowledge spatially and are situation-bound, i.e. context bound information (DeWalt, 1994: Murdoch and Clark, 1994). In contrast to science, local knowledge occurs in some cases in the form of magic, and it lacks its own philosophical expression or knowledge (Knight, 1980). Further, differences between the two forms of knowledge exist primarily in the methods of knowledge appropriation/ extraction and in the nature of the resources employed in the practical application of the acquired knowledge.

Therefore, a look at the "mechanics" of the intermingling of various knowledge systems and the complex relations among them in situations of cultural contact as seen in the course of history may be necessary in future. However, at this juncture it is sufficient to say that "Both are attempts to make sense of the world, to render it comprehensible to the human mind. Both are based on observations and on generalizations deriving from those observations" (Berkes *et al*, 1995). Researchers today can more easily correlate nature of local soils and plants with the rural people. Present knowledge also facilitates soil surveys and mapping in a few days which would otherwise take months (Howes, 1980). In this regard, Abdel Ghaffar Ahmed (1994) states that neither indigenous nor scientific knowledge should be over glorified. Instead, the right mix of the two knowledge systems can help to enhance the prospects for sustainable food production in rural areas. It is therefore necessary to consider local knowledge as the base on which to plan sustainable development. "Problems of resource management relate not only to physical planning and efficient deployment of modern input delivery systems, but also includes the local communities and their readiness to accommodate change" (Salih, 1992). What is required is a two-way flow of communication between modern technologists and ordinary peasant farmers. "The least that can be said is an idea borrowed from the people, developed by the agronomist and returned to the people again is much more likely to be adopted than something totally alien to the culture" (Richards, 1975).

Indigenous Knowledge as a Theoretical Approach in Development

Over the last decade local knowledge has emerged as an applied theoretical input for socioeconomic planning. Theory of cultural ecology and neo-evolutionism ideologies stem from the influence of environment on ethnological development. The basic features include the impact of environment on culture of local groups with focus on adaptation, and reciprocal links between culture and ecology as is seen by indigenous groups who rely on forest resources for survival (Barrett, 1996).

Indigenous conservationism is now a widely accepted ideology that is culturally expressed in conservation ethics, and in natural resources management and animistic religious beliefs. It is noted during field observation that indigenous conservation has been identified as a major thrust area for many indigenous groups who continue to rely heavily on natural resources for their subsistence economies.

It has also been observed in the past decade that indigenous knowledge is often incorporated into agricultural systems and participatory development when non-governmental organizations and outside agencies implement development projects. Often, when government involvement or stakeholders in highly valued natural resources are involved, indigenous knowledge is often overlooked and ignored.

By contrast it is also observed that in the name of development, many countries have attempted to force indigenous groups to abandon their traditional livelihoods in order to "save the forests" and assimilate them into modern society. Many such development and conservation related projects have failed due to lack of involvement and knowledge of those local people. In the past, some top-down approaches were also initiated by outside agencies attempting to develop natural resource management or extraction techniques, which were also unsuccessful due to local ecosystem constraints (Gegeo, 1998) and lack of local perspectives. The adoption of indigenous knowledge has on the other hand led to a locally informed perspective, which transformed itself into development. The most relevant approach to working with indigenous groups is the neopopulist one where the focus is to contribute long-term positive change, by promoting culturally appropriate and environmentally sustainable adaptations as local resources are commercially exploited (Sillitoe, 1998). Indigenous participation is used as a process of empowerment to amplify traditionally unacknowledged voices (Slocum et al., 1995). Participation focuses on ways to mobilize local resources, engage diverse social groups in decision-making, and identify patterns to eliminate poverty (Feldstein and Jiggis, 1994: Slocum *et al.*, 1995). Active and meaningful involvement of indigenous people with the researchers (outside agency) regarding issues related to land tenure and social-cultural development, and in relevant decision making process a partnership of equals while linking ecological and social justice concerns. It is important that researchers (or agencies) link management practices to social and environmental ambience of particular communities where such practices are present (Eghenter, 2000).

Review of Literature

Numerous studies have been conducted by eminent scholars and experts on indigenous knowledge system and natural resource management. Some of these studies have been reviewed for the purpose of this research so as to provide panoramic view of the indigenous knowledge system and its trends across the globe.

'Indigenous knowledge' is a very broad term as discussed earlier touching all aspects of life of the indigenous people who through their own experience have been able to accumulate extremely rich knowledge of their own environments. It is the knowledge pertaining to weather, land, rivers and seas, of various breeds of animals and plants.

According to Meehan 1980 such studies of indigenous knowledge about natural resources which has been carried out over the last five decades or so falls into two broad categories: ethno-science and human ecology. For him, ethno-science refers to the local knowledge systems that relate broadly to biological phenomena comprising a number of sub-fields like ethno-botany, ethnozoology, ethno-medicine, ethno-veterinary, and so on. Human ecology draws on the biological science systems to account for human beings' relations with their environments, considering them to occupy niches like other animals which they have to exploit, interfacing with their various technological assemblages (Orlove, 1980).

Ecological knowledge as studied by ethnoecology (an approach that focuses on the conceptions of ecological relationships held by a people or a culture), may be considered a subset of ethnoscience (folk science), defined by Hardesty (1977) as "the study of systems of knowledge developed by a given culture to classify the objects, activities, and events of its universe." Pioneering work by Conklin (1957) and others documented that traditional peoples in Philippines often possessed exceptionally detailed knowledge of local plants and animals and their natural history, recognizing as many as 1,600 plant species. Scientific experts acknowledged indigenous environmental knowledge e.g., Arctic ecologist Pruitt used Inuit (Eskimo) terminology for various types of snow for decades.

Similarly there has been growing recognition of the capabilities of ancient agriculturalists and water engineers and architects (Fathy, 1986) have developed appreciation of ethnoscience, ancient and contemporary, which gradually lead to the acceptability of the validity of traditional knowledge in these fields.

Ancient ways of knowing began to receive currency in several disciplines, including ecology. Posey (1985) has shown that many indigenous groups in diverse geographical areas from the Arctic to the Amazon have their own systems of managing resources. Thus, the feasibility of applying TEK to contemporary resource management problems in various parts of the world is being gradually recognized.

According to WCED 1987, tribal and indigenous peoples'... lifestyles can offer modern societies many lessons in the management of resources in complex forest, mountain and dry land ecosystem as the local communities are the repositories of vast accumulations of traditional knowledge and experience that link humanity with its ancient origins. Their disappearance will be a loss for the larger society, which could learn a great deal from the traditional skills. Gadgil and Berkes (1991) highlight that traditional knowledge is relevant for managêment of contemporary natural resource such as wetlands. "Rules of thumb" developed by ancient resource managers and enforced by social and cultural means, are in many ways as good as Western scientific prescriptions.

Johannes (1989) expressed that people who are dependent on local resources for their livelihood are often able to assess the true costs and benefits of development better than any evaluator coming from the outside. Their time-tested, in-depth knowledge of the local area is, in any case, an essential part of any impact assessment.

The use of traditional knowledge may therefore benefit development agencies in providing more realistic evaluations of environment, natural resources and production systems. Involvement of the local people in the planning process thereby improving the chance of success of development programmes (Warren et al., 1993).

In the past, modern science alone provided biological and ecological insights, the knowledge base for resource management, conservation, development planning and environmental assessment. At this stage of the development of TEK, it is possible to say that indigenous peoples and the knowledge held by them do have something to contribute to each of the above areas. But traditional knowledge is complementary to western science, not a replacement for it (Knudtson and Suzuki, 1992).

Nowadays, the economically developed world as well as some of the underdeveloped economies still depend on their own and the world's forests for their survival, yet in different ways and meanings (Seeland, 1997).

Ahmed (1994) highlighted that developing countries were forced to export natural resources like timber, oil and other minerals, and export crops in order to pay for the import of modern technology and inputs. The disparity between developed and developing countries has hampered the attempt to build technology on the basis of indigenous practices and values. Omari (1990) opined that indigenous belief systems have been destroyed by the introduction of a money economy with its capitalist mode of production, the introduction of religions like Christianity and Islam, and by the introduction of state control of natural resources. He further argued that western value systems have had negative effects on traditional African land ethics. Being influenced by the new values, people now use natural resources as objects for exploitation and profit making.

According to Kebede and Hurni (1992), in many circumstances real participation by farmers is not feasible because of their obvious lack of knowledge about the processes and about the means used by outsiders for positive intervention. Government officials also view peasants as the victim and the cause of environmental disaster. However, a significant player in the land degradation process, one which is conveniently forgotten, was the government itself" (Dessalegn, 1997). The land tenure system during the imperial era discouraged peasants to make long-term investments on the land. When they improved the quality of the land and raised the yield of their harvest, the landlords increased tenancy rents (Cohen, 1987).

Governments in other "Third World" countries have become a hindrance to the development of indigenous environmental knowledge. They feel that participatory projects and the technology development research which involves peasants are politically sensitive and dangerous (Thrupp, 1989). Attfield and Belsey (1994) stated that the protection of the natural environment was neglected in Western civilization. Although the philosophy of nature is at least as old as the pre-Socratics, the natural environment has not been given due attention until recently. The main concern of moral, political and social philosophy has been the social environment rather than the natural environment. Likewise, anthropologists restricted their study to subjects such as languages, religions, social structures, arts and crafts by ignoring ecological relations between human beings and their environment (Klemm, 1985).

Conklin (1957) opined that the failure of development projects have drawn the attention of many international organizations to reconsider the potentialities of indigenous environmental knowledge in development planning. There is a large number of literature that recognizes the role of indigenous environmental knowledge all over the world. For instance, Conklin reports that adults in the Hanunoo tribe in the Philippines identified 450 animal types and 1600 plant types. He adds, "partly as a result of this great interest in plant domestication and detailed knowledge of minute vegetative differences, native categories outnumber by more than 400 types, the taxonomic species into which the same local flora is grouped by systematic botanists.

Clarke (1977) study highlighted that the Bomagai-Angoiang people in Papua New Guinea are classic shifting cultivators who use "palaeotechnic" rather than "neotechnic" agricultural strategy. They use time-bound products and conditions on a small scale as an aid to self-maintenance within the absorptive capacity of the existing ecosystem. They employed poly-cultural diversity as effective agricultural strategy. Clarke is of the opinion that it would be useful to combine the advantages of primitive group structure with those of modern technology in order to improve the status of the world environment. He suggests that the immediate task of humanity is to apply the principles of permanence and to build into the neotechnic—dominated ecosphere a structure of permanence. If we do not, we will all—neotechnic and palaeotechnic alike-become victims of progress.

Some indigenous knowledge is not only useful on terrestrial environments, but also in the sea. Among others, Pacific islanders depend on their knowledge of the behavior and movements of the marine animals much of which is new to science (Johannes, 1989). Bennet and Cross (1993) stated that oral traditions and indigenous knowledge are still important among some groups in Europe including the mountain Protestants in France and Italy, the Gaelic-speaking Scottish islanders.

The African people have a considerable knowledge of their environment and rational forms of land management because they are close observers of weeds, pests, crop conditions, soil types, weather and environmental changes (Richards, 1975: 1989: Johannes, 1989: Chambers, 1989: Omari, 1990: Salih, 1992: Salau, 1992: Medani Mohamed Ahmed, 1994: Abdel Ghaffar Ahmed, 1994: Yagoub Abdella Mohammed, 1994: Ayers, 1995). African peasants have a unique system of soil and water management. Various ethnic groups in the semi-arid zone of Western Africa adapted traditional soil, water and fertility conservation strategies to ecology and human diversity. On the contrary, modern administrators imposed the modern strategies of management on the people, without taking local ecology and human diversity into account (Roose, 1992).

Yagoub Abdella Mohamed (1994) reported that the harvesting of sheet flow run-off water is useful in the central Sudan. In the same way, Babiker (1996) studied the indigenous techniques of water harvesting in Dar Hamar, Western Sudan, which are dispersion of farming units, dry planting and mixed or intercropping in response to the unpredictability and unreliability of rainfall, temporal and spatial variability. The Hamar tribe store rain water "in the hollowed out trunks of the baobab trees, in shallow and deep hand dug wells, natural ponds and large depressions" (Babiker, 1996).

Some studies also show that the Oromo people give important value to the natural environment (Knutsson, 1967: Bartels, 1990: Hultin, 1987: 1994). Kassam and Gemetchu (1994) argue that the Oromo have some of the finest principles and codes of behavior towards nature. They underscore that the Oromo maintain a perfect balance between nature and culture. They further outlined how the Boran pastoralists protect the natural vegetation and manage pasturelands through a combination of different mechanisms.

Some scholars convincingly indicated the danger of idealizing village communities and their knowledge. Not all knowledge and activities of local people are valid and environmentally sound. Some of their practices have had undesirable local environmental effects. "The value of traditional and environmental knowledge and management practices thus cannot be taken for granted. Some relevant traditional beliefs are incorrect or misdirected. Superstitions sometimes over-ride objective observations" (Johannes 1989: Klemm, 1985).

For thousands of years the American Indians have survived in the Amazonia, thanks to their understanding of ecological zones, plant/human/animal relationships, and systems for the management of natural resources, which are the product of complex processes of experimentation (Posey: 1991). Posey (1985:1991) studies shows that Kayap' Indians have in-depth knowledge of such subjects an ethnoecology, ethnobotany, and ethnozoology. They see forest management as an integrated system of plant communities, rather than individual species; similarly, wildlife and even semi-domesticated bees are part of their overall management strategies.

The discussion on the role of local institutions in natural resource management is a recurrent theme in relevant literatures on southern Africa. The goals of these institutions are to play a significant role in sustainable development, particularly in poverty alleviation, empowerment of the rural population, and natural resource management. They are important custodians of indigenous knowledge systems and culture, and also intermediaries between the Government and the communities. Thus, the degradation of these institutions will result in the degradation of indigenous knowledge systems as there will be no inter-generational transmission of such knowledge (Mutowanyika, 2000).

Steward's (1955) theory basically talks about the cultures that have evolved within their local environments, unlike general theories propounded by earlier scholars, and its in-depth analysis of the relationships between specific cultural institutions and their local environments shows how and why those cultural institutions evolved and persisted. In his analysis Steward (*ibid*) found that some cultural features are also more crucial to a society's utilization of the resources and some other environmental features are essential for subsistence activities than others. Briefly those features would constitute the cultural core, which are closely related to subsistence activities and economic arrangement. This is evident from his analysis of the evolution of the patrilineal band and argued that similar forms of social organization have developed in societies living in very different environmental conditions, as a result of the fact that they hunt similar types of game.

In his approach he emphasized the primacy of behavioural considerations in subsistence technology and the extent to which other cultural features are affected by those behaviour patterns. In his analysis he was concerned with the operation of a particular variable in relation to a limited set of variables, not in relation to an entire social system.

In fact after Chambers (1979) drew attention to the importance of local knowledge, no reference has ever been made to ethnoforestry in particular by any research scholar. Dr. Klaus Seeland (1997) who is working on local knowledge on forests has a similar view on the issue.

According to Gadgil, Berkes and Folke (1993) Biodiversity Conservation practices are as diverse as the cultural diversity in the world. Indigenous knowledge of uses of local plants, animals, habitat preference, life-history and resource availability is socially transmitted from one individual to another within and across generations, in the same manner, though not necessarily in writing, unlike scientific knowledge. In addition, there are examples where communities regulate the use of resource by restricting the access to resources, and enforcing compliance through religious belief, ritual and social convention. It is debatable whether these 'restraints' evolved after trial and error or as systematic prescriptions. However, it is certain that these restraints definitely contributed to the cause of biodiversity conservation.

Sacred trees symbolize specific arrays of human conditions, possibilities and anticipation. In India, species of trees are worshipped as manifestations of gods, as representatives of particular stars and planets, and as symbols of the natural elements-energy, water, land, air-each of which has its own independent and relational meanings. As sources of social cohesion, continuity, and control, religious trees shape human actions (Chandrakanth and Romm 1991).

Kapila Vatsyayan (1992) informs the sacredness of trees seen in every part of India. Deodara (*Cedrus deodara*) is considered the abode of the gods; Sal (*Shorea robusta*) is venerated in Uttar Pradesh, Bihar and Madhya Pradesh. Similarly Rudraksha (*Elaeocarpus* sp), Bel (*Aegle marmelos*), Ashok (*Saraca asoka*), Kadam (*Anthocephalus chinensis*) and Pipal (*Ficus religiosa*) are considered sacred in Rajasthan (Singh and Pandey 1982).

Sacred groves, found in many parts of Asia, Africa and many Island States, are one of the most important forms of protective ethnoforestry. They are also the center of cultural and religious life of people living in villages located near the groves. Sacred groves are located near a temple or village deity.

Origin of sacred groves can be attributed to slash and burn system' of agriculture, where several forest patches were left unfelled around farmlands. These groves came to be institutionalised as centers for culture and religious life. This is one of the prudent use patterns of natural resources. It is extremely difficult to arrive at a workable prescription on quantity, quota, or protection to different life history stages that decisively guard against resource depletion. Providing refugia such as sacred groves may then be easily perceived as most efficient way of guarding against resource depletion (Gadgil and Vartak, 1976).

A study from Andhra Pradesh found a total of 761 sacred groves of which 361 were investigated for floral wealth. Of these, 161 are of timber value, 413 of medicinal importance, and 278 yield non-forest produce (Prakash Rao's article in WWF-India Newsletter).

Das (1997) describes sacred groves to be a worldwide phenomenon. In islands of Phillipines and Malaysia certain trees are cultivated for their fruits. Tribes also perform ceremonies to enhance growth. Jena (1997) reported several sacred complexes in Orissa and there groves are called by various names such as Jahira by the Hos, Jaher by the Santals, Jhanka by the Dudh and Dhelki Kharia, Sarana by the Munda tribe.

To restore the sacred groves of Aravalli hills in India, a programme, Aravalli Sacred Grove Conservation, was launched in 1992. This programme includes protection of existing groves, planting of indigenous species, soil and water conservation and participatory approach for restoration (Pandey, 1994b).

In India, the Rajasthan Forest Department is relaying on both scientific and indigenous knowledge of forest ecosystems to formulate their management strategies to combat the problem of deforestation (Pandey, 1996). Indian societies at all stages of socio-economic and cultural evolution exhibited practices of restrained resource uses that may have the consequence of prudent use of natural resources (Gadgil and Guha, 1992). Recently a workshop was held on local knowledge of forests to discuss the issue (Seeland and Schmithusen, 1997). Presentations focussed primarily not on ethnoforestry but on societal issues and ethnobotany of tribes in Orissa State in India. We are not aware of any other direct reference to Ethnoforestry, published in any of the journals of repute. Another indication of the knowledge gap is the fact that, to the best of our knowledge not any of the many reviews on ethnobotany has mentioned or explored the specific domains of ethnoforestry.

M.G. Chandrakanth and Jeff Romm (1991), without explicitly referring to ethnoforestry, examined the likely policy implications of religious forest classifications. Using material primarily from India, they described religious systems of forest classifications that often coexist with government's classification in legal, economic, and ecological nomenclatures.

The studies carried out by (Sinha, 1967: Mann, 1968: Guha, 1985: Wade, 1987: Nadakarni *et. al.*, 1989: Nadakarni, 1990), highlighted the social, economic, cultural and religious aspects involved in the traditional management of common property resources of the community. They eventually said that these forms of community management witnessed a decline after the end of the 19th century and this reduction was attributed to population pressure, state intervention and operation of market forces.

Impacts of deforestation on the cultural values system are clearly delineated by (Roy, 1990: Sarkar, 1990). Different cultural mechanisms established by the society to overcome the problems of handling resources during the scarce situations is provided by (Mann's, 1968: Sinha's, 1967) works on the Ladhaki and the Birhor tribals respectively. Works done by Shankaran, 1992: Berreman, 1977: Prakash Reddy, 1982) also highlighted the role of cultural factors involved in the coping mechanisms and adaptive strategies. How the religious ethos were playing a significant role in preserving the sacred grooves have been discussed by (Lynn White, 1959: Gadgil and Vartak,1990).

Most important critique to the common property resource theory was made by the anthropologists, who pointed out that the communally owned resources can be effectively and efficiently managed as it was proved by the different studies of (Ciriacy-Wantrup and Bishop, 1975: McCay and Acheson, 1987: Erasmus, 1977: Ottenberg, 1966: Ostrom, 1987: Netting, 1972: Berks, 1985: Anderson, 1987: Pinkerton, 1987: Kurein and Vijayan, 1995). They all pointed out that in simple societies people have effectively generated institutions and rules to limit the over exploitation rates.

Contemporary ecological movements, especially Chipko as well as other conflicts, over natural resources by fringe communities have become a matter of great concern in social science research in India (Vandana Shiva, 1991: Gadgil and Guha, 1996: Baviskar, 1995: Kurien, 1993: Berreman, 1989: Jain, 1984: Omvedt, 1987). These studies focus mainly on the role of social and cultural institutions in the preservation of natural resources.

Kothari (1998) has amply demonstrated how different practices of management of resources exist in the communities that live close to the forests physically as well as culturally. In the book "Communities and Conservation: Natural Resource Management in South and central Asia", communities are defined as those people who derive benefit from the forests directly or indirectly; not as a particular ethnic group of people defined by their caste or religion.

Ramana Rao (2000) in a study on the tribals and the ecological change in Mahadevpur in Karimnagar district of Andhra Pradesh found that that state forests have a variety of minor forest produce derived from a number of species of plants, besides a large number of animals and minerals. Most of these areas are rich in forest produce. The forests contribute considerably to the economy of the tribals. The tribals in this region largely make their living through the collection of Minor Forest Produce in the forest areas. Their indigenous practices of forest management ensure sustainable availability of these produce.

Jodha, N. S, (1986) points out how resource use in an economy is socio-culturally dependent. For example, in India we have a majority of rural population, which depends on common property resources for their livelihood. He has found in his study that while 84% of the rural population depends on common property resources for food, 100% of them depend on these resources for their fuel. Resource management practices, he asserts, ensure efficient and sustainable supply of these resources to the local people.

According to Madhav Gadgil (1991), the roles of custodians of biodiversity management should be bestowed to local communities of people who live close to the earth, i.e. tribals, peasants, herders, fishermen and rural artisans. These are the people who depend for their day-to-day survival on the biological resources of their immediate surroundings. They have for centuries obtained livelihood without destroying the natural diversity, its rapid erosion is largely due to large-scale commodification of nature in the last two centuries. They have an intimate knowledge of natural resources, in their very limited localities. Experience of recent years has shown that communities of such people, also called the ecosystem people by Dasmann (1988) are often willing or even eager to take on the role of custodians of local natural resources.

Ramakrishan, (1992) observed that resource management practices among tribals are deeply interwoven in their day-to-day lifestyle, and forest related livelihood pursuits. Their understanding of the environment is also reflected in their philosophy, which embodies a host of traditional belief and practices and acts as a cultural means of conservation. Numerous examples of environmental conservation through such traditional beliefs and practices can be seen among tribal societies of India.

Maikhuri *et al.*, (2000) stated that the Bhotiyas of the Central Himalayas, who practice migratory cattle-raising and traditional agriculture, are highly dependent on the resources which they find in nature. Authors documented their knowledge of medicinal plants and wild edibles, and the specific manner in which they are used. This information is presented in three tables. The article ends with a discussion of the need to conserve this knowledge in the light of the rapid acculturation now taking place within the above referred subcommunities viz Tolchha, Marchha and Jadh.

Bandyopadhyay and Saha (1998) bring out the farmers' ingenuity in devising ways to obtain sufficient seeds under extremely difficult conditions. As there is no organized system of seed supply in the extremely remote, humid and tropical Andaman and Nicobar Islands Territory in India, individual farmers have developed techniques for selecting and preserving the seeds of the most important food crops.

Jain and Sneh Lata (1996) draw the comparisons between various communities within a country, or between communities located in different regions. They stress on great scope for establishing the credibility of certain groups' unique knowledge, as well as for expanding its application. Authors also present medicinal uses of plants, which are found in both India and Amazonia. Some of these uses are unique to indigenous communities of Amazonia. The present study opens up the prospect of their increased utilization in the Indian region.

Studies among the Konda Reddi

The earlier accounts that we have of the Konda Reddi are from Hemingway (1905) and Thurston (1909). Hemingway, during British administration, refers to the Konda Reddi in, the Gazetteer of Godavari district as, a caste of jungle men. In his account, he observed that Konda Reddi speak a corrupt form of Telugu and call themselves Pandava Reddi, Raja Reddi. Thurston writes that the Konda Reddi, "live by shifting cultivation". Beef eating is taboo, while pork eating is very common among them. They worship the spirit of hills, they are superstitious, and believe firmly in sorcery and call wizards at the time of illness. The most comprehensive study of the Konda Reddi is Haimendorf's, "The Reddis of Bison Hills: A Study of Acculturation" (1945). This study threw some light on their beliefs concerning human sacrifice. In his study he mentioned that they are still at the Neolithic stage with pre-agricultural techniques. Misra (1998) focuses on the relationship between Konda Reddis and plants. In his paper he contemplates on the nexus between Reddis and the trees around their habitat, as perceived by them and manifested in their cosmology, worldview, religion, ritual and folksongs, which is contrary to the Western notion of dualism between culture and nature. Sunita Reddy (2004) attempts to view Konda Reddi from the ecosystem approach. The study gives an emic perspective, the local perception, and gives the indigenous people a voice. Misra (2005) documents the Konda Reddi knowledge of forest and forest produce, wildlife, agriculture, animal husbandry, and ethno-veterinary practices, ethno-medicine, insects and flies, food and food reserves, etc. in their present form. The author also throws light on natural resource management and common property resources. An attempt has also been made to relate indigenous knowledge with resource management and development among them.

From the above selected reviews it is noted that attempts have been frequently made in different areas to understand the patterns of indigenous knowledge and the domain of management of natural resources by the local people. The researchers have also tried to explain the conditions and the processes of collective choice, patterns of resources uses, and changes in it due to external influence in the course of history. Notwithstanding the conventional bias towards the superiority of western science to local indigenous knowledge, it is certain that the recognition of local knowledge gained significant ground when the application of scientific knowledge in various development programmes failed to bear the intended result due to social and cultural obstructions in the acceptance of rational knowledge, which sometimes stood in contradiction to the local knowledge and belief of the local people. Despite the scholastic contribution of all these studies, it is felt that the study of local knowledge system is by and large removed from the total cultural context in which it is located. Sometimes while local knowledge is construed as superstitious and an impediment to the process of development in less developed countries at other times there is an unusual clamour in favour of local knowledge system as if that is the ultimate panacea for the survival of the forest resources of the communities. The majority of tribes in India are indigenous and continue to practice use of indigenous knowledge in their day-to-day activities, which is largely undocumented. The indigenous knowledge system National Programme of India is targeting mainly this knowledge system used by "tribes" in remote areas of India which have been largely neglected, relegated to the periphery and therefore remain largely unprotected. In general, India lacks an enabling and protective policy environment for this particular knowledge system.

Therefore, there is an urgent need to dispassionately enquire and honestly study the local knowledge and resource management practices of the local communities, for which this research has been taken up. Local environmental knowledge and natural resource management, for the purpose of this study, is basically understood as the description and analysis of the local flora and fauna as understood by the community under study.

In this direction micro-level study plays a significant role leading to a comprehensive understanding of a particular area, or village. These studies are more useful in policy formulations as well as for evolving culture specific resource utilization patterns for diversified regions. In this direction, to have a comprehensive understanding of the local resource management process, the present research mainly emphasizes on some aspects of "local environment knowledge and natural resource management among the Konda Reddi of Andhra Pradesh".

Conceptual Framework

It has been argued that indigenous knowledge is related to the culture of a group of people, including its identity and spiritual and religious beliefs. While some scientists and development planners may see traditional knowledge as a means of resolving certain development problems, indigenous people see it a part of their overall culture, vital to their survival as peoples (Dewes, 1993).

Recent emphasis on empirical research, scientific literature and documentation on indigenous environmental knowledge reflects the increasing awareness about traditional ecological knowledge. The magnitude of the taxonomic system suggests that the extent of traditional knowledge may be quite profound, and that, indeed, taxonomy is important (as in the biological sciences) as the basis for building extensive systems of knowing about nature.

In the recent social science literature, traditional knowledge to a large extent has been documented, especially, for a large number of so-called foraging peoples, from the tropics to the Arctic (Johannes, 1991). Such traditional ecological knowledge has been found to have management relevance, especially in regard to sustainable use of renewable resources (Freeman *et al.*, 1991). It is evident that the need for traditional knowledge research has emerged within cognitive anthropology more recently by reestablishing consciously or unconsciously, the particularistic paradigm proposed by renowned American Anthropologist Franz Boas, who was the first exponent of cultural relativism. He explains human culture as the bundle of traditions in a certain geographic region. To him, all cultures are unique and particular in nature and there is no need to dichotomize culture either superior or inferior. The contemporary participatory approaches to development planning are increasingly stimulating the significance of traditional knowledge. In the domain of participatory development planning, anthropologists attempt to examine the depth and breadth of indigenous/traditional knowledge based on ethno-science methodology for discovering the cognitive aspects of people. It rested upon an emic point of view and requires understanding of the symbols and meaning attached to natural phenomena.

The thrust of our argument is that 'development from below' is a more productive approach than that from 'above' and an essential ingredient is indigenous knowledge. Cognitive anthropology stresses upon eliciting indigenous knowledge as a base for development planning.

The ecological dimension to traditional systems of knowledge is usually taken to refer to the people's medicinal, technical, and ritual uses of plants, animals and rocks; to place names and occupancy of territory; or to the spiritual, cosmological, and relational aspects to the various presences (animate, inanimate, present or past) in an environment.

From the multidimensional perspectives, we can define traditional environmental knowledge as a body of beliefs transmitted through oral tradition and first-hand observation. It includes a system of classification, a set of empirical observations about the local environment, and a system of self-management that governs resource uses. Ecological aspects are closely tied to social and spiritual aspects of the knowledge system. The quality of traditional environmental knowledge varies among community members, depending on gender, age, social status, intellectual capability, and profession (hunter, spiritual leader, healer, etc.), with its roots firmly in the past, traditional environmental knowledge is both cumulative and dynamic, building upon the experience of earlier generations and adapting to the new technological and socioeconomic changes of the present.

The local knowledge system perspective when looked along with the management of natural resource perspective, offers well-sustained and embedded approach for resource management, which help in sustenance of the society for a long time through various cultural upheavals. In this study I have adopted an anthropological approach to holistic perspective to understand the interaction patterns of the community living in the village with their natural resources.

Statement of the Problem and Purpose of the Study

Although indigenous peoples constitute one of the largest vulnerable segments in contemporary society, they and their knowledge systems have been marginalized mainly because of the craze for modernity and globalization. The distinct culture of indigenous people and their identity, their economic activities, religious beliefs, notions, and traditional ways of managing natural resources are often regarded as backward and superstitious and absolutely incompatible with modern society and development.

Indigenous knowledge has often been dismissed as unsystematic and incapable of meeting the rapid economic growth needs of the modern world. Historically, modern societies have regarded indigenous traditions as less progressive, and as a result many groups of indigenous peoples, especially their younger generations, tended to devalue their native cultures and to adopt new lifestyles and technologies. Consequently, IKS could not be captured and stored in a systematic way and are therefore endangered. The lust for modernity and new technologies is threatening the great store of knowledge held by indigenous people. A good number of indigenous groups in India and elsewhere in the world have suffered from long-term discrimination, inequity and exclusion from the planning and execution of development programmes and projects.

Despite the serious erosion of IKS over the decades in many communities in India, they are still relevant for promoting sustainable development of the country. Far from being anachronistic in the contemporary world, these knowledge systems have much to offer development policy-makers, environmental managers, administrators, and stakeholders.

Today, indigenous knowledge systems are at risk of becoming extinct because of rapidly changing natural environments and fast pacing of economic, political, and cultural changes on a global scale. Practices vanish, as they become inappropriate for new challenges. However, many practices disappear only because of the intrusion of foreign technologies or development concepts that promise gains or solutions to local problems. The tragedy of the impending disappearance of indigenous knowledge is most obvious to those who have developed it and make a living through it. But the implication for others can be detrimental as well, when skills, technologies, artifacts, problem solving strategies and expertise are lost.

Today, much local knowledge is being lost and, along with it, valuable knowledge about ways of living. Moreover, Indigenous knowledge is not yet fully utilized in the development process. In this backdrop, the present study is an attempt to record Konda Reddis extensive knowledge on agriculture, forest and wildlife resource management with an emic approach.

Objectives of the Study

The specific objectives of this research are as follows:

(1) To delineate an ethnographic account of the culture and habitat of the Konda Reddi with regard to various aspects of economic and social life;

(2) To explore the local knowledge of agriculture and utilization of natural resources by the local community;

(3) To understand the native conception and perception of environmental knowledge of the community regarding classification and quantification of the flora, fauna and land domains;

(4) To delineate the local knowledge system development and conservation with respect to sacred groves, and the institutional arrangements in managing them, including common property resources.

Research Methodology

Anthropologists have always investigated not only the behaviour and the material products of people, but also their thoughts. In the anthropology of the last thirty years, studies on the so-called emic view or emic perspective have gained a certain predominance. For these reasons, information on local knowledge can be found in studies from all branches of current anthropology. Having said that, these are mostly case descriptions emphasizing the unique character, the cultural specificity of the studied knowledge (Brush, 1993). Considering its importance, anthropological research tools and techniques have been used for the present study to capture the holistic understanding of the local knowledge system.

Selection of the Study area and Village

The selection of the village was based on the following factors (1) The village, which has wide coverage of forest, the forest existing in the village ecosystem that is collectively owned and managed, from which economic and social benefit is being derived by the society. (2) It was then decided to select a community, which comes under the Primitive Tribal group (PTG) category, which is practicing concept of sacred groves belief system, and still living in some form of relative remoteness. This was possible, only with the community that still believes and depends on its local knowledge system which is useful for sustenance or which is still relatively cut off from the mainstream because of inaccessibility of the region due to lack of proper communication system. The Konda Reddi tribe has all the above said features required for the present study. Therefore, Konda Reddi was selected for the present research. The Konda Reddi constitute the largest Scheduled Tribe in the district of East Godavari in the state of Andhra Pradesh, and are largely concentrated in the Addateegala Mandal of the district. Konda Reddi, predominantly live in this area and pursue their livelihood directly depending on forest resources. For field work Addateegala Mandal of the District was selected. The reason for selecting this Mandal was that this region has a mix of agricultural and forest economy for its population, which provides an opportunity to study the various resource management patterns existing in the community.

The East Godavari district is covered with variety of soils and climate; it has variety of forests with different species. The area of forest in the District is 3,197 Sq.Km. constituting 29.58% of forestland. The whole of the agency tract in East Godavari District lies in the Eastern Ghats and the same is named as *Rampa* agency. The Eastern Ghats rise by gradations from the level of the coast and spread throughout the agency. The district is connected by train route. The National Highway 07 runs through the district. The tribal population in East Godavari is highly spread over seven Mandals of the district. The Konda Reddi's relative seclusion and unspoilt tradiation, ethnicity made their village the ideal choice for studying local knowledge systems and management practices in relation to the natural resources in the area. Though the villages in the region were heterogeneous in nature in nature, Konda Reddi dominate in number in most of the villages. This enabled me to study the presence/penetration of knowledge system in other communities in a village, where the natural resources are managed collectively.

Ducharthi Ramavaram Panchayathi is located in Addateegala Mandal which has high concentration of Konda Reddi population. Most of the villages are close to the forest as well as Naxal infested. It became then imperative to choose a village, which would be closer to the forest, yet not too far from a main road, so that I could be evacuated in case of an emergency. With these backdrops, from the above Panchayathi Utalapalem was selected for this study. The study village is about 22 km. from the Mandal Headquaters. The distance between the village and forest is less than one km.

Methods, Tools and Techniques

The present study adopted participant observation as an important research tool. This method professes to study of a culture in its totality. In order to study the above-mentioned aims and objectives, it was essential to do fieldwork in the region by interacting with the people by listening and talking to them. For same, data were also collected from the secondary sources on demographic profile, physical features, forest conditions and types etc., from the respective departments like Mandal revenue office, forest department, and tribal welfare department etc. The fieldwork was carried out to collect the data from the study area. The duration of field was spread out in two phases (May-December 2002, March-June 2003). The total duration of the fieldwork was about one year.

I have used various techniques for eliciting information pertaining to Konda Reddi and their local knowledge perception regarding various natural resources and soci-cultural milieu. As Pelto and Pelto (1978) states, "if it is our serious purpose to understand the thoughts of people, the whole analysis of experience must be based on their concepts, not ours."

The present study is basically a qualitative micro-level study aimed at understanding the local environmental knowledge among the Konda Reddi. The requisite fieldwork was undertaken in the village, *Utlapalem*, under Addateegala Mandal. In order to fulfill the objectives of the study, qualitative and quantitative anthropological tools and techniques have been used. These are mainly observation, participant observation, interviews, focus group discussion, guided field walk, case study, schedules, using detailed checklist and key informants.

Household Schedule was prepared for collection of data on general information at the household level. The schedule consisted of questions ranging from family details to cultural practices of the community; the community political system, religious system, economic system, social system, land holdings patterns, occupational patterns, status of common property resources, natural resources etc., and interaction with the outside world.

The first phase of the present research was aimed at identifying the locations, cultural framework, range and variety of the various plants and animal species. Information was gathered on culturally significant plants, which had an obvious, dominant place in the social life of the Konda Reddi. This phase was contextual in that no direct questions were asked about what plants people knew or used. Observation was the key approach here, since the researcher wanted to learn about the specific cultural sensibility concerned with plants. This was important because many people did not at first divulge what plants were used for healing rites and other ritual purposes. Most of the information in this phase was gathered during one month of field research.

The second phase consisted in identifying and interviewing key informants or specialists on land, plants and animal species. During this phase the aim was to study attitudes towards land, plants and animals local knowledge and norms for use of natural resources. The indigenous experts were asked to think of information about natural resources which I would verify at a later.

The specific data in the second phase regarding the local knowledge of natural resources was gathered through interviews with the key informants in the village associated in the management of natural resources; like *Goravadu* (priest), *Bandroth* (village messenger), and village head man (*Kulam Pedda*), *Vaidudu* (medicine men), who are the custodians of information on the cultural practices. The information on sacred groves with the village social system, rituals, worldview, taboos and prohibitions etc were gathered from them besides the key informants and people. I actively participated participated in *Gangalamma, Bairudu*, birth, puberty, marriage, and death ceremonies to know the symbolic explanation of the ritual process and meaning the community tries to derive from some occurrences. Discussions with the people of the village form the basis of data on the management of common natural resources, the matter of ownership and utilization rights of the people of the village, differential usage, forest produce, grazing, etc.

From close observation I could determine the behavioural pattern amongst the family members, likewise the relations between parents and children, husband and wife, sibling and socialization of a Reddi child. Observation is an important tool used in walk abouts visiting sites where I expected to see plants in use from forest or display, such as at home and market places. Anything that may be relevant to the subject being investigated is noted, and where necessary, questions were asked for further clarification. Wherever possible, I would take some one along from the village to the forest and fields, who could provide the necessary explanations.

In addition, I also participated in some activities like helping in drying minor forest produce, collecting harvested paddy, winnowing paddy, carrying paddy to their home. Other activities included hunting trips during important religious ceremonies and participating in rituals, chanting during the ceremonies. Through participant observation I could get a better understanding of cultural processes involved in indigenous resource management, their use and distribution patterns. I have also used non-participant or quasi-participant observation in gathering information on specificities like technique of grazing, milking goats, etc.

Interview as a tool was used for both groups and individuals. Informal interviews were held in the market places, on the way and at other gatherings (toddy place, cock fight places, play ground, hunting spot etc.). Where as, more formal interviews were held in community work-spots and meetings, or prearranged household visits. With the help of interviews, I could gather information on ethno-biology such as plants that are used for food (fruits, vegetables, green leaves, mushrooms, tubers, roots, shoots, etc.), plants used for house construction, agricultural implements, forest implements, plants used for curing human and animal diseases, plants used as fibre, fodder, dye, etc.

In-depth interviews were used for collecting information on local agricultural knowledge, shifting cultivation, traditional crops, development of new crops, pest control, ethno-biological classification, nutrition and food processing, kitchen garden plants, developing of new farming systems and other general plants and animals uses.

Group discussion method was used to collect data on farming practices, leadership structures and decision-making patterns, health practices and delivery systems, traditional medicines, labour sharing arrangements, local indicators of poverty and socio-economic standing, indigenous taxonomic schemes, and other related information. Focus group discussion method was used for gaining a better understanding of the knowledge about socio-cultural resource management practices and barriers in the process. An intensive discussion with the villagers was held to draw resource maps of the village.

Author's Field Experience

Before leaving for my field area, I had to read some literature involving theoretical premises characterizing the field of study, published literature in numerous book, journals and magazines. The idea was to acquaint myself with the community where I was going to carr out my intensive research work. The village I intended to work is the remote area of East Godavari District. Being the naxalite prone area the opportunity to work in such place seemed to be receding away from my hands. But my inherent prediction overshadowed the mettle spirit within me and I developed courage to work under any crucial circumstances.

During my first visit to village, I could only succeed firstly in established my rapport with the villagers. Secondly, capturing rough idea regarding the location of the place. I initiated my fieldwork but it seemed to be lagging in every respect I couldn't talk to the villagers fluently due to the difference in dialects. In my first visit it was me who was kept under observation throughout my stay, my movements inside the village were being discussed by the villagers. But this hardly had an impact on my work. Being already cognizant about the belief of villagers that the anthropologist is thought some kind of spy that he is gathering information for the purpose harmful to the community. With the help of few informants especially youth, I got to know some of the stories being circulated about me.

First I tried to get close with the youths and children of the village. For this I started going to the school, which nevertheless seemed to be a mere playground for the children rather than a school. I tried my best to change the scenario of the place and convinced the children to come for the purpose of study. But, it was not as easy as I had thought to change the mindset of children and their parents. They always found it useless due to lack of study bent. Steadily by giving my own example I tried to corner the children whom I found interested and initiated to teach them. Initially handful of students turned up and responded positively. But soon my perseverance started showing magical results. The students developed an interest at least to come to school regularly. Besides social work, I felt comfortable in grabbing the information, though not so much regarding my work. Now the villagers even began to take keen interest in me. I was placed at heart, as my work didn't go unrewarded. I used to play volleyball and cricket with the youths.

Although, I tried my best to accept and reciprocate friendly but couldn't get the same response from the natives quickly.

The whole village community was indifferent towards the hygiene. Hardly any one of them has the notion of taking bath properly. Though I could not do it in a day, as my intervention could have spoiled my rapport entirely. As soon as I joined the school, I took a special class of hygiene and health stated making them cognizant of cleanliness.

Problems of everyday pragmatics include how to program one's time in the field, when to get up, when to eat, when to sleep, when to release, when to cook, when to allow informants to visit and when to write up the field notes. The following passage describing my fieldwork in Utalapalem highlights some of the problems of everyday pragmatics. As the main occupation of the villager is agriculture, so whole day they used to spend the time in the fields and refused categorically to talk. At night they seemed to be too tired, relaxed themselves with toddy 'Tati Kallu' (energizer for them), thus to overcome fatigue. Thus again they were not in a position to talk. So I had to gather information during early morning or while helping them during fields. Being an anthropologist I had to keep patience till I got well adapted and my work started gaining rhythm. The denizens of the village accepted me, which made my work quite simplified. I could helve every information from the people, which initially seemed to be the cumbersome job.

The main success behind the work of anthropologist in how well he blends with the local community. For this he has to develop a mindset to break down the strict resolution of life. His clemency to some extent him flexible to work in any environment. As far as anthropologists are concerned this is the most crucial stage where they need to compromise. I even had a face such situations oftenly.

The local drink toddy, which they called 'Tati Kallu', is usually served to everybody on their great festivals. As to understand their rituals, agricultural seasons etc. I had to attend such ceremonies. I was served this 'Kallu' every time as it is considered a good omen. Being aware of the fact that it had some sedative effects, I could not deny them, as it would have been regarded as breach of honour. So, I had to accept it without any signs of disgust. Hence, I closely feel that this attitude of an anthropologist drive him closer to the destination and he/she can reveal every aspect of his data from the natives.

During my first visit I furnished to gather information pertaining to agriculture, forest plants, wild animals, utility of other resources like cattle, land, water in the village. But still in depth information were not substantiated. For data collection I used to go forest with villagers and also few times went to hunting with them. I could only spend six months during my first visit. In the 1st trip I choose to stay in an old Aganwadi School rather to get accommodated in anyone's house. Thus my next visit was inevitable.

The inception of my next visit to the same landmark began next year. This time I was fully equipped with the tools of knowledge, rapport build was already to some extent and moreover there was trust and confidence among the people. I was whole-heartedly accepted in the village without in under hustles, which I had to face in 1st trip. I felt as if I was also the part and parcel of the village community. Being in cognition with every part of village, this time I choose to stay in a headman' house rather to get accommodated in old Anganwadi school. But still the problem of toilets arose. I had to construct a toilet with Borassus palm leaves and bamboo. I felt embarrassed every time. I had to enter it. Somehow I managed in it for some days. But it was a temporary one and couldn't last long. While being in my field it had blown away by wind. Now I had to walk outside especially during nights.

Now, I was the known person of the village, the people used to come up freely with their problems and I had to listen to them patiently. Sometimes they even expected me to write applications of complaints etc. I had to indulge in lot of community work. I used to go weekly markets along with them. I used to take the sick to nearby hospitals. Sometimes they woke me up during midnights. I never made them feel that I am an alien. I blended with the whole community as a person is attached to his family.

However, I also participated in some activities like going to field and forest along with them, helping them in their daily activities, like carrying and drying minor forest produce, collecting harvested grains, making meals sometimes and taking care of their children mostly when the agricultural activities kept mothers out of their houses for longer periods.

This approach helped one in each and every step of my data collection. The people felt free while talking to me and the sense of hesitation vanished completely. The people used to escort me whosoever I required to collect the information. Second field visit was a real cakewalk for me.

This time I stayed for six months and returned in the month June. People now had become so emotional so did I. They persisted me to stay and construct a house in the village itself. It was easy to say than to do. I had my own limitations as I cajoled them. When I caught the bus I found everyone in tears. This was really painful. Some of them accompanied me till Kakinada. The next year people came to meet me in Hyderabad. And I had to fetch them to my hostel. Still my field memories are fresh there in my mind and they will stay there forever. Still today village people call me at least once in month and they discussed village matters and other things. After my P.hD. field work I have visited the village twice. They were very happy when I visited them. Village people always invite me on their celebration of festivals and ceremonies.

Organization of the Book

The present study is dealt in seven chapters. The first chapter is composed of introduction to the subject, including review of literature, conceptual framework, objectives of the study, selection of the community and the study village, research methodology and significance of the study.

The second chapter consists of ethnographic account of the culture and habitat of the Konda Reddi with regard to various aspects of economic and social life.

The third chapter explores the Konda Reddi knowledge of agriculture. The chapter attempts to provide information on *Podu* cultivation, dry and wetland cultivation and how local people categorize and perceive their environment.

The fourth chapter describes folk perceptions of local environmental knowledge; the way people describe their physical environment and forms of utilization of natural resources among the Konda Reddi.

The fifth chapter describes how they name, classify, quantify the fauna. Further the chapter discusses the*Konda Reddis* knowledge of wildlife. It includes their general characteristics and behaviours, and hunting practices.

The sixth chapter is about the local practices of natural resource management. The chapter provides discussion on sacred groves from common resources perspective, and the benefits the society avails from it, the value it associates with it and importance of it in the overall dynamic from micro to macro level.

The last chapter consists of summary and conclusion.

2

Ethnographic Profile of the Study Village

This study was conducted in the village Utlapalem. Utlapalem is a small Konda Reddi village located in the Addateegala Mandal of East Godavari District in Andhra Pradesh. The major tribes inhabiting the East Godavari district are Konda Reddi, Koya, Konda Dora, Konda Kammara, Valmiki and others. Among the major tribal groups inhabiting the Integrated Tribal Development Agencies (ITDA) area of Rampachodavaram, the Konda Reddi tribe has been recognized by Government of India as a Primitive Tribal Group (PTG) since the year 1980. The Konda Reddi inhabit the banks situated on either side of the river Godavari in the hilly and forest tracts of East Godavari, West Godavari and Khammam districts of Andhra Pradesh.

District Profile

East Godavari district is situated on the eastern bank of the river Godavari. It is bound in the North by the Visakhapatnam district and the state of Orissa, in the east and the South by the Bay of Bengal and in the West by Khammam and West Godavari districts. The district is situated between $6^{0}30''$ and

18°30" N latitude and 81°30" and 82°36" E longitude. East Godavari district can be broadly divided into three physical zones viz, the agency area or the hilly tracts, the upland, and the delta. The Eastern Ghats rise by gradation from the level of the sea-coast and spread throughout the agency tract. The costal belt is blessed with highly fertile deltaic soil, giving it the reputation of being the rice granary of the country. The upland area, though less fertile, supports various varieties of dryland and rainfed agriculture. Most of the agency area is however hilly and covered with dense forests.

The district comprises 57 Mandals attached to 5 revenue divisions. The district covers an area of 10807 sq. km with a total population of 48,72,622 of which the tribal population is 1,76,086 i.e. 3.61 per cent of the total population (P.T.G. Action plan, 2000). The Konda Reddi population as per 1991 Census is 76,391.

The total forest area in East Godavari District is 2,850 sq km. and this works out to 26.4 per cent of the total geographical area of the district. Three fourths of the forest area is hilly and the rest is in the plains. The forests are mostly South Indian deciduous type. The geographical area of the agency belt is 4629.31 sq. km, which forms 42.8 per cent of the total geographical area of the district.

Climate

The climate differs considerably from that of the more central parts of the Deccan. The area is subjected to both the Southwest and Southeast monsoons. The hot season lasts from the beginning of March to the middle of June. The district rainfall is 1,160 m.m. There is hardly any rain in March; the rainfall is between 3 and 4 m.m. in April-May and about 40-45 m.m. during the Southwest monsoons, which lasts from June to September. According to records about 9 m.m. rainfall occurs between October and December.

Soil

The soil of the district is mostly black, red loam, and laterite, one of low fertility. This is deficit in organic matter, nitrogen and lime. Konda Reddis are aware of the huge deposits of precious and semi-precious stones in the rocky and mountainous places near their villages. Because of administrative restrictions, Reddis do not venture out to mine them.

Flora

The Konda Reddi are hill people inhabiting the forests. The forests are mostly deciduous type. There are *Mamidi chettu* (Mango tree), *Tati cheetu* (Borassus Palm), *Vepa chettu* (Neem Tree), *Shikakai Chettu* (Soap nuts) *Chinta Chettu* (Tamarind Tree), *Artocarpus integrefolicus* (panasa), cashew nut, *Jeerugu/ Jeelugu*, (*Aeschynomene indica/Caryota* Tree), *Annona squamosal* (Sitaphalam), *Papaya, Jama Chettu* (Gauva Tree), Teku Chettu (*Tectona Grandis*), *Karakkayala Chettu* (Myrobalan Tree), *Usiri Chettu* (Goose berry Tree), *Nimma Chettu* (Lemon Tree), *Veduru* (Bamboo), *Rai chettu,* Maredi chettu, *Nerodi chettu, Regu pandlu, Parmi chettu, Palachettu.* They also grow flower-bearing plants like *Bantipuvvu* (Merigold flower), *Gannerupuvvu* (Oleander flower), *Mandarampuvvu* (Hibiscus flower), *Kanakabarampuvvu, Chamantipuvvu,* etc. in their front yards.

Fauna

The fauna includes wild animals like tigers, leopards, bears, jackals, deer, wild dogs and foxes, which are occasionally found. Small animals like wild pigs, black buck, monkeys, nilgai, jungle sheep, and wild cats are commonly seen in the jungle. Commonly found birds include water birds of many descriptions, cranes, storks, pelicans, herons, egrets, gulls, kingfishers and other birds like peacock, pigeon, parrot, goose, dove, mynah, jungle fowl, hornbills, etc. Domestic animals include cows, oxen, goats, buffaloes, poultry, dogs, cats etc.

Village under Study

The study village is located on a hill by the name Pathapadu. Due to lack of water facility, the village has moved down the forest and is called Utlapalem. The villagers explained that the tribes were leading a nomadic life in the early days. They started looking for permanent settlement on observing others.

As narrated by the villagers, the name of the village Utlapalem is derived from the words "Utalu" and "Palem" as "Utalu" refers to streams, and "Palem" refers to village, as people settled in this village gradually. Subsequently, the settlement was named Utlapalem. After some years the village was officially recognized by the government. The stream is located near the village on the eastern side about half a km from the village. The villagers used the stream as a source of drinking water. The village itself was named after the stream. Konda Reddi say that the legendary leader Alluri Sitharama Raju, who valiantly led a tribal revolt against the British, dug up this stream. Stream water is available throughout the year. The source lies in hills called *Nilaramma Konda* located 2 kms away on the western side of the village. *Gangalamma*, the village deity of Konda Reddi represents the source of water. Most of the Reddi villages are found near the sources of stream water, and therefore, water and streams are intimately associated with their religious as well as secular life.

General Description of the Study Village

The study village Utlapalem is about 22 kms away from the Mandal Headquarters. The village Utlapalem is part of D.RamaVaram Panchayati, which is 4 kms away from the study village. The Mandal comprises 99 villages. The village is connected to neighbouring village Anukulapalem by a 2 kms road. To reach the village Utlapalem, one has to get down at the Anukulapalem village on the main road and walk about a kilometer and a half towards the north on a muddy road.

Utlapalem village is situated very close to the hilly forest. The total population of the village is 326 with a male population of 165 and female population of 161. Utlapalem comprises 57 households. The village shows scattered arrangement of houses. Out of the 57 houses only two houses are semi *pucca* and the remaining houses are *kacha*.

The structures vary from square to rectangular, built with mud walls and *Borassus palm* thatched roofs. Each house is independent of the other, surrounded by courtyards, and scattered. Usually, there is a Verandah in front of the house, and at the back are two rooms invariably separated by a mud wall. The courtyard/Verandah is often used for cooking; sometimes a separate kitchen hut is constructed on the east side of house. There is a *kacha* road, which connects to the main road of the Mandal. The village has a primary school and the school building is located at the corner of the village in the South direction. The study village has 3 major sources of water. The first one is a stream (*Uta Neeru*) near the village, the second source of water is bore wells and the village has five bore wells used for domestic purpose. All the bore wells are functional, but one among them is not used, as the water from that pump has a foul smell. The third source of water is a canal. Water from this canal is used for agriculture in rainy season, washing clothes and cattle, and for ablutions. The villagers prefer stream water for drinking as it tastes good and is available throughout the year. In summer they use the stream water for various purposes. At the center of the village, is located lord Rama temple (*Ramalayam*). It is also a common meeting place for the villagers.

Physical Appearance

Renowned anthropologist Christopher Von Furor-Haimendrof characterized Konda Reddi as small, sturdy men, with legs that are short compared to the length of the body. The skin colour varies from a very light copper brown to dark chocolate, but medium brown with ruddy and sometimes yellowish tinge may be considered the average skin colour. They have heart-shaped faces with small pointed chins, high and prominent cheekbones and flat noses with wide nostrils. In the interior hills of *Maredumilli* and Addateegala blocks Reddis are generally distinguished by extreme coarseness of features with taller and heavier builds. The hair is wavy. Facial hair is negligible and men have a sparse growth of beard and moustaches. In spite of the divergence in the racial features, the Konda Reddi are easily recognizeable (Haimendrof, 1945).

Dress Pattern and Ornaments

The men folk are usually clad in dhoti, lunges and banyans, shirts, long turban. Men wear dhoti, which comes up to their knees and is called *Pancha* in their local term. While at work men wear *Gochi* a lion cloth to cover their waist. But most Reddis wear trousers and shirts while going to the *Santha* (shandy) or visiting other villages. Whereas, the women's attire is *saree*, and blouses in which one end of the *saree* is taken over shoulder so as to cover the chest. The sari is worn both with front pleats and back pleats. Generally, young women adopt the former and older women the latter.

It is commonly noticed that Konda Reddi women wear nose rings (*Mukkupuduka*), earrings (*Chevikammalu*), anklets (*Pattilu/ Gajjalu*), necklace (*Medagolusu*), stone chains (*Nallapusala/ Dandalu*) and bangles (*Gajulu*) etc. Married women wear silver toe rings (*Mattalu*), green bangles (*pachagajulu*) and gold or copper maglasutra (*taali*) and beads (*Pusaladanda*) around neck, which identify them as married. Men wear silver or copper finger rings; wearing earrings has become old fashioned.

At present elderly men are generally seen with *Siga*, a kind of hairstyle that means a half-shaven head and hair that is tied in a knot near the occipital region. With the contact of plains people, the young Reddis are habituated to the modern haircut in the saloons. Womens hair is plaited or braided in coils and matted. Women tie their hair in a knot near the occipital region. Women are fond of flowers and decorate their hair with the available flowers. Both men and women use coconut oil to dress their hair.

These days, traditional dresses are worn during customary rites and festival only. Still traditional dress can be seen among village folks. With all the modern market facilities around, they are opting for synthetic fabrics. As a result of urban contact, youth are wearing modern dresses like jeans, trousers and T-shirts etc. some of the affluent tribal women wear gold ornaments. Among the younger generation girls, ornaments made out of cheap metals have become popular. Transformation can particularly be seen at the weekly market places. The way of dressing among the Reddis has also changed. Besides their weekly rations, they purchase cosmetics and new clothes. Consumerism among *Reddis* has increased due to the flow of goods to the markets through non- tribals.

Material Culture

The tribals of the village possess items like cycles, watch, radio, tape recorders etc. Recently the villagers purchased a mike set and DVD player to watch movies on weekends. The following Table shows the material possession of the study village.

Table 1
Total Number of Material Possessed in the Present Study Village

Name of the village	*Cycles*	*Watches*	*Radio*	*Television*	*Torch Light*	*Cots*	*Bullock Cart*	*Sprayer*
Utlapalem	35	38	32	2	25	59	19	06

The utensils used for domestic purpose are mostly made of stainless steel, aluminium, brass and bell metal. Vessels made of stainless steel; aluminium and earthen pots are mostly used in the study village. Different sizes and shapes of vessels are used in every household. A pair of steel/earthen (*Bindelu*) is found in every household and used for fetching water.

Women use various water containers to fetch water like *Bindellu* (it is a vessel made by bronze, steel, peethal), pots *(Kundalu)*, Buckets, *Dabbalu* (tin and bronze vessel) etc. *Gollem* is a big container for storing water, which is made of mud etc. Men fetch water in a yoke (*Kavedi*). It is made of bamboo stick with vessel holders made of ropes hanging on two ends of the bamboo stick. Filled water containers are placed in these rope holders and the *Kavedi* is placed in a balance on the shoulder and carried. This helps the Konda Reddi to carry more water from far off places with less strain.

Drinking water is stored in *Bindellu*, and for other multiple purposes, they store water in buckets, *Gollem, Dabbalu, Ginnelu, Kundalu*, etc. *KondaReddis* store water in earthen pots in summers, which keep the water cool for drinking. Generally, almost all the house members fetch the water in the house but women do most of the fetching.

For grinding small quantities of the grain they used traditional grinder (*thirgalli)*. Large quantity of grains takes to grind in rice mill. They use furniture like chairs, wooden and fibre, wooden cots, bamboo mats, wooden Table etc. For storing the food grains bamboo and cane baskets are used.

Demography of the Study Village

The 57 households are divided into 6 lineages. However no lineage-wise pattern of residential distribution is observed. The following table shows the distribution of village population by gender.

Table 2
Distribution of Village Population by Gender

Name the of lineage	*Male*	*Female*	*Total Population*
Pallala	42	41	83
Rolupalli	46	35	81
Bachala	10	11	21
Mamidi	12	06	18
Janmuri	11	11	22
Jartha	09	07	16
Total	130	111	241

It is evident from the above table that the total village population is 241 out of which the male is 130 constituting 53.94% of the total population and female population is 111 constituting 46.5% of the total population.

Types of Family

The family constitutes the basic unit of the village social structure. It functions as the social and economic unit in a society. It is fundamentally a domestic unit among the Konda Reddi, and is called '*Kutumbam*' in their local language. Family is an institution that legitimizes procreation apart from performing certain other social functions. The organization of the family depends on kinship ties, so that every individual in the society is associated with his or her family. The family plays a very important role in both individual as well as group activities. Among Reddis the family unit is characterized by sharing of a common hearth by all the members. Further, each family functions as a single economic unit for production as well as consumption. Sometimes, two families might produce crops jointly, but they divide the crops after harvest. The Konda Reddis family is patriarchal and patrilocal. Surname (*Intiperu*) and property (*Asthi*) of the person is passed on through male line only.

Among the Konda Reddi family, males are dominating. Any matter related to family materializes only after the decision by the male head of the family. In the village nuclear families are dominant. Most of the village people think that joint families create more problems. Every individual establishes a family of his own immediately after marriage. After marriage a woman moves from her natal family to her affinal family.

The eldest male member of the family functions as the head of the family. Generally, decisions are made only in consultation with other members of the family. Women are considered as social and economic assets in the Reddi society. They play a significant role in the socialization of children and also in providing labour for earning a livelihood for the family by attending to all kinds of subsistence activities. It is observed that in the present village under study nuclear, extended, and joint family types of families exist. The following Table shows the lineage wise distribution of family types.

Table 3
Lineage Wise Distribution of Family Type

Lineage name	*Nuclear family*	*Extended family*	*Broken family*	*Total House hold*
Pallala	13	03	02	18
Rolu-palli	09	05	04	18
Bachala	04	01	00	05
Mamidi	02	03	00	05
Janu-muri	02	03	01	06
Jartha	04	01	00	05
Total	34	16	07	57

It is evident from the above table that nuclear families are 34 in number. The nuclear family is of the predominant type, constituting 59.64% of the total families. The extended families in the village are 16 constituting 28.07% of the total families.

There are 7 broken families constituting 12.28% of the total families in the village.

Types of Marriage

Konda Reddi are an endogamous tribe. This is to ensure parity and uniqueness of their tribe and also to stop infiltration of non-Konda Reddi. Marriage among Reddis is both an intra-village as well as an inter-village affair. Marriages depend on lineage segments and these are divided into many subgroups with individual family names. After marriage, the bride gets affiliated to her husband's lineage. Practice of exogamy is prevalent for regulating matrimonial relations. This is evident from the fact that people do not marry among their blood relatives. Marriage by negotiations, by love and elopement, by service, by capture, by exchange are socially accepted ways of acquiring mates. Levirate is common.

Generally the villagers seek marriage alliances with the nearby villages. Nowadays their marriage alliances are spread over a radius of 25-50 kms. For example the following villages *Veerampalem, Tungamaduguala, Panasalapalem, Yerlagadda Y.Ramavaram, sheihrai, Chintakarapalem,* fall under the above radius. The villages falling under this radius are preferred villages for establishing marital relations. Father's sister's daughter is given more importance for marriage alliances than mother's brother's daughter.

Monogamy is a predominant feature of the village. Cross cousin marriage (*Menarikam*) is the most preferred form of marriage although marriage by negotiation and marriage by mutual consent and elopement are widely practiced by the Reddis. Only in very rare cases, marriage by capture or service takes place. It is observed that marriage by negotiation is regarded as the ideal method, whereas the more practical methods of acquiring mates are marriage by mutual consent and elopement. In case of elopement, social acceptance is given only after the groom gives a sumptuous feast to all the villagers. Marriage by negotiation is an expensive process, which involves more time, rituals and consequently money. Divorce is very rare and remarriage is quite prevalent. There is no dowry system. Bride prices are nominal. The proposal for separation through socially accepted divorce could be initiated by either of the partners. Though they believe that marriage is socially important, the stability of the marriage is not guaranteed. *Kulamkattu* (village council is considered as the socially approved body to execute divorces. The village council decides compensation and other formalities by which marriage is legally dissolved.

According to Hindu custom the ideal age of marriage for boys is 21 and for girls is 18 years. In the early days, among the Reddis, a large number of girls were married off even before they attained puberty or immediately after puberty. But in the study village, the boys are married off between 18 and 21 years. The girls are married off between 16 and 18 years.

Among the *Konda* Reddi, marriage payment is in the form of *oli/kanyasulkam* (bride price), a payment from the family of the groom to the family of bride, which can be either money or items like grocery etc, The payment of *Oli/Voli* has a very significant place in the marriage proceedings. Unless the *Oli/Voli* is paid, the marriage is neither socially accepted by the bridegroom's parents nor the village elders. *Oli/Voli* (bride price) under which the prospective groom would have to pay some cash. The amount depends on the status of the groom. The groom's side will pay cash, give two bags of rice and other eatables, including meat for the party, which follows the marriage function. The expenditure for the marriage is borne by the bridegroom's family and it differs from one family to another depending upon the standard of living. The bride brings her belongings and households' things as discussed before the marriage.

Reddi men and women have the liberty to select their life partner. Like men, women too have the right to divorce and remarry but this custom is gradually declining with the contact of urban societies. The marriage ceremony is held at the groom's marriage *Pandal*. Marriage ritual starts by giving a ritual bath to the bridge and the groom. During the time of marriage, women sing their traditional songs related to the marriage occasion. If a married woman abandons her husband

and elopes with another man, the man with whom she eloped should pay her first husband bride price money and all the marriage expenses; it is compulsory to host a feast for all the village elders as penalty.

Among the Reddi, dowry system was prohibited in the early days. However, there are instances where the practice of *Oli/Voli* system is reversed and some people in the community have started demanding dowry. Due to influence of non-tribals they started accepting the dowry system in their social structure. This indicates the changing attitude of people towards wealth. With this intention, people are raising their economic resources. Moreover, grooms' parents also are demanding property like land, cattle, goats, etc. Due to the impact of Hindu model of celebrating marriages and functions, expensive weddings have become common in the community. Trapped by the outside consumerist culture, the Reddi are imitating the non-tribals.

Marriage with Non-Tribal People

There have been instances of anti-social elements taking advantage of the socio-economic situation of tribals to lure away the tribals from the tribal area, destroying their culture. Some government officers, contractors and businessmen and other outsiders obtain the free company of the young Reddi girls for a few rupees. These young girls become mothers before marriage and their children do not even know who their father is. Irrespective of the economic position of Reddi families, expensive weddings have become common among the Reddi.

This is a new trend emerging from contact with non-tribals. Konda Reddi are allowing their daughters and sons to marry non-tribal. It is observed that a few Reddi women and men are married to non-tribals. These non-tribals are maintaining shops, which were allotted to their tribal wives by the government.

Kinship Terminology

The Konda Reddi do not have a language of their own. They speak a dialect of Telugu and their accent is slightly different from pure Telugu. The terminology of village people is influenced by Telugu terminology. They follow the Telugu kinship terminology. An individual calls his father as *Ayya/ Nana* and his mother as *Amma/Awwa*. He addresses his wife by either referring by one of their children's name or by calling out her name. He calls his son by his name or as *babu* and his daughter as *pilla* or *bujji*. The elder son is usually addressed by his name or as *pedda babu*. The younger sibling calls his/ her brother as *Anna* and elder sister as *Appa*. Younger brother is addressed either by name or *chinnanna*. Younger sister is called *chellai*. Elder sister's husband is called *bava*. Younger sister's husband is called as *bammardi*. Elder brother's wife is called as *vodina*. Younger brother's wife, mother's brother's daughter and father's sister's daughter are called *maradalu*. Formal relationship is maintained between father-in-law and daughter-in-law, son-in-law and mother-in-law and also in case of husband's elder brother. This ensures mutual respect and restraint. It strengthens the marital bond and at the same time minimizes conflicts.

At the other extreme is a teasing relationship. It is a particular combination of friendliness and antagonism. By teasing each other and playing games, they develop a friendly relationship, which protects them from conflict and hostility. In the village, relationship exists with elder's brother's wife and younger brother of her husband, brother-in-law and sister-in-law, nephew and aunt, etc. relationship leads to friendship and it is meant to enhance the relationship.

Economic Organisation

The economy of the Konda Reddi is mainly categorized as subsistence economy based on shifting (*Podu*) dry and wet cultivation. Agriculture forms the most important occupation of the Reddi as does the collection of minor forest produce such as different varieties of leaves, tubers, roots and fruits etc for their day to day consumption. They sell these items to the Girijan Cooperative Corporation (G.C.C.) or sometimes in emergency they sell to petty shopkeepers in *Santha* (shandies or weekly tribal market) or to the brokers of local business men. Though fishing and hunting are practised they are not for commercial purposes but for domestic consumption. It is a subsistence economy, which implies self sufficiency for most part of the year but not through out. These people also get bank loans to purchase bullocks, goats etc. Their main assets are land and livestock.

Generally they cultivate paddy in wetlands and other crops along with vegetables in dry land. Income from commercial crops like *Saggudumpa* (tuber), cotton, tobacco, and sale of vegetables help them in supplementing their day-to-day family income. Besides, working as wage labour is another subsidiary means of earning cash income for a few villagers during off-season. Villagers consider cultivation as the main source of livelihood, which provides food and work throughout the year. Apart from this, the villagers are engaged in basket making, carpentry, black smith, domestic animal rearing, hunting and fishing and also in the collection of minor forest products.

In addition they collect *beedi* leaves in summer for a month. The leaves are collected and arranged in bundles of 100 leaves each. The average price of the bundle is approximately Rs. 13/- but the buyer pays the amount in accordance with the quality and quantity of the bundle.

The village Vana Samarakshana Samiti (VSS) setup by the Forest Department of Govt of Andhra Pradesh includes a few of the villagers as its office bearers with the advice of the forest department officials. The VSS committee recruits both males and females aged between 15 and 50 years as daily wage labour. The VSS workers have to work in the plantation in the specified areas popularly know as VS Coop i.e. in the local forest area. Each individual who works in the VS Coop get Rs. 20/- per day and this work is available for two months in summer. The villagers also avail themselves of this opportunity.

In the study village all the households are in a state of indebtedness because they do not have enough income and surplus resource to maintain their families. In emergency they get money from moneylenders. Generally, people borrow money for agriculture, treatment of diseases, starting business, construction of house, expenditure on children's education and sometimes for seeking employment.

In the village, half the people have taken money from the government as loan. Taking money from the government involves a long process and it is difficult for them to get money when they are in need. Besides the government does not take any risk to supply credit to the village people and it has been neglected by the agencies. The indebted village people are not able to pay back, as their income is not sufficient for their daily expenses. But they repay the moneylenders by selling their goods to them.

Livestock Rearing

The animals reared by KondaReddi are cows, oxen, goats, poultry and others such as cats, dogs etc. in their villages. The particulars of the livestock in the village are given below.

Table 4
Particulars of the Cattle and Others Livestock of Village

Oxen	*Cows*	*Goats*	*Poultry*	*Calf*	*Dogs*	*Buffaloes*
128	56	89	189	32	35	3

The Reddi rear cattle primarily for the purpose of agriculture and secondarily as source of making some money. In addition to this it is useful for bridegroom parents as bride price/dowry at the time of marriage as a kind of marriage payment. And also during the time of festivals/rituals/ ceremonies the cattle is used for the purpose of sacrifice.

Livestock rearing is part and parcel of the Konda Reddi agricultural economy because it not only assists them in various agricultural operations such as, tilling of soil, transportation of agricultural produce etc., but also provides basic inputs in the form of organic manure and basic source of meat protein. It has also become a means for earning cash income for villagers as the demand for animal and poultry products has increased and market facilities for selling the products have also emerged. They use the manure of cattle for their cultivation. They keep the livestock in *Godlapaka* enclosures to their houses.

Political Organization

In Utlapalem village, both traditional and modern political organizations exist side by side. The study village Utlapalem comes under D. Ramavaram Grampanchayat and Yellavaram constituency of East Godavari District. The traditional village authority council is called the (*Kulam Kattu/Kulam peddala koluvu/Panchayat*). Both the inter-tribal and intra tribal disputes are settled in this council. This council acts as the supreme legal, social and political body of the village and the village head man (*Urupedda/peddakapu*), acts as the judge of the tribal council. The traditional village authority council consists of the village headman (*Urupedda/peddakapu*), *Pujari* (*Goravadu*), secretary and treasurer. The headman of the village heads the *KulamKattu* in the village now called a member. In the absence of the village headman the next member is performs the duty.

Generally the role of village authority council is to look after the village problems and development activities in the village. It discusses civil, criminal as well as ritual matters relating to the village and takes appropriate decisions. The council also discusses matters relating to development activities in the village and takes decisions. In the village authority's meeting every one has the right to express their views and urge for the justice. The village authority takes the decision or passes its judgment after hearing the complaint and explanation from both the parties. The decision of the village authority is ultimate and cannot be questioned. It is observed that every individual wishes to settle matters of dispute in the purview of village authority. But the final decision-maker is the village headman who makes the decision based on the collective opinion of the village. Village headman enjoys social prestige and is placed first in the local hierarchy.

The violations of social norms prescribed by the society are taken seriously and are punished accordingly. If the culprit fails to pay the penalty sometimes he may be ex-communicated from the community on the basis of the seriousness of the offence. Generally, the punishment for any guilt is in the form of a fine in terms of rupees and animals which is paid to the authority council.

The tribal council discusses and takes decisions on major socio-religious and customary problems of the tribe. For example celebrating village festivals such as *Gangalamma, Kappapanduga, Kothamasa, Bhumipanduga* etc., which are to be initiated on auspicious days. Village headman conducts meeting of the villagers in initiating the programme in the village. And decisions will be taken with consent of villagers.

It takes up issues like family quarrels, elopement, divorce, adultery and similar other incidents, that would degrade the prestige of the whole village and the tribe. If any individual violates the social norms or violates the decision of the *KulamKattu*, it is considered a serious offence and is punished accordingly.

A Gram Panchayat selects one representative called village member in each village to assist him and advice him in the matter of dispute settlement. The assistant is selected in consultation with the village members and in the absence of village headman the assistant adjudicates the matter on his behalf. Most of the time the assistant Gram Panchayat, along with other village functionaries like *Urupedda* (headman), *Goravadu* (*Pujari*), village member and *Bandrothu* (village messenger) participate in the Gram Panchayat. All adult male and women members of the village under its jurisdiction attend the Gram Panchayat meeting.

Along with their traditional village authority council, the villagers also take active part in the modern political system. The villager either elects or nominates individuals to represent the Panchayat as the members of Panchayat and V.S.S members, and the DWCRA members. All these members address the problems that the villagers face in their respective political bodies and sort them out. Though they do not have any special status in *Kulam Kattu*, they have a special say in the village meeting. The people of the village are aware of the election process and they have been voting for the past 30 years. The people are also sympathetic to the naxalites and police, who supposedly come and meet them often. For the selection of other functionaries, election is held in the village under the guidance of the officer. Election is conducted every five years.

Due to impact of the modern political system, villagers do not follow the hereditary system for the selection of village headman. It is reported and observed in the village that every individual wishes to settle matters of disputes in the village authority council itself and are extremely reluctant to approach the police station or any legal court for their settlement. If the matter is very serious and people are against the village authority council than only they approach to police station.

Today, the community elders or headman do not have more powers. The power or authority they enjoyed through traditional Panchayats or village council is gradually evaporating. Their role has been changed with the increase of interaction of the community with the bureaucracy and political parties and other non-governmental agencies. Villagers exposed to the outside world have their cohesion breaking up. There is a gap between the educated youth and community elders in understanding the problems and decision-making process. The traditional authority of the elders in the village has weakened, due to salary-earning tribal leaders influenced by non-tribals.

Social Organization

Kinship ties among the Konda Reddi are very intense. Every man and woman is a member of the village community and of its lineage and these two units, together with the family form the main pillars of the Konda Reddi society. Through out the tribe, the functions of village community are fundamentally the same. The village amounts to 57 households, which are concentrated in one place. Lineage members are considered as descendents of a common ancestor and are known by a common name. This is known as the '*intiperu*' and precedes a man's personal name. Konda Reddi have adopted *Intiperulu* (surnames) to regulate their marital relations. *Intiperulu* of the lineage members are considered as the testimony that they are the descendants of common ancestors and hence, are known by a common surname. Descent is strictly patrilineal and the marriage ritual includes a particular ceremony, by which bride is transferred from her own to her husband's lineage. The number of the Reddi family names is so great that the complete enumeration cannot be attempted. But there are only 6 family names that could be found in Utlapalem village i.e. *Pallala, Mamidi, Jartha,* Rolu*palli, Bachala, Janumuri.*

Cooperation in the production and the gathering of food is one of the main links of social cohesion within the family and to a lesser degree within the village community. But it is through family, that major part of adaptation and production activities are carried out. The aspect of mutual exchange of labour is the significant mechanism which encourages service among inter and intra lineage members in the village. This aspect of mutual exchange of labour pattern marks the *Konda* Reddi group solidarity and corporate activities.

Religion

Reddis consider the religious rites as designed to propitiate the supernatural powers on whose their wellbeing depends. Emotional reactions accompany with the worship of various deities Reddi worship a Gods and Goddesses, like *Gangalammatalli, Bhumidevata, Bhairudu, Parentalu, Kondadevatha and Adavirajulu*, etc. Among the *Reddis* religion acts as a medium that transmits animistic beliefs and practices to the generations that follow. *Kondadevatha or Adavirajulu, Bhairudu* are forest Gods residing in the forest. They appease Gods and Goddesses by celebrating rituals/festivals in order to get a good harvest and to be free from disease and other misfortunes. In *Utalapalem* village each and every agricultural activity begins only after worshiping or celebrating a festival/ritual of the concerned deity. That is starting from sowing of seeds, to harvesting the crops. These festivals are mango harvest (*Mamidikotha/Korrakotha*), pulses harvest (*Pappupanduga/ Chikkudukotha*), paddy harvest (*Varikotha*) etc. These are the harvest rituals with an offering to the God of the particular crop harvested. These are community offered rituals and they do not consume the crop until it is ceremonially offered to the deities.

Apart from these, they celebrate many other festivals like *Gangalamma Panduga* (village deity festival), *Bhudevi panduga* (goddess of earth), *Vannadevudu panduga/Kappa panduga* (god of rain), *Kondadevatha/Adavirajulu panduga* (God of forest), *Bhairudu panduga* (God of domesticated animals) etc. During the celebration of these festivals they worship with great devotion and reverence. They also sacrifice animals and poultry in the name of the village Goddesses and Gods *Gangalamma, Chintalamma, Muthyalamma, Nillaramma, Durgamma, Saralamma, Gautamma, Bhudevi, Pandavulu.*

They believe in many ancestral spirits, and ancestor worship (*Parentalu*) is an important feature of the religion among Reddi. It is believed that the spirits of the dead have to be satisfied. Violation of this norm may result in the wrath of the spirit, which would harm the people and the settlement as a whole. Therefore *Reddis* perform rituals to appease these spirits. These festivals have been performed and transmitted from the time of their ancestors. Reddi also believe that the family members after their death are reborn (*Puttagathulu*) into the same family due to love and affection towards the living members. Therefore, *Reddis* name the newly born children after the name of their ancestors, preferably their grand parents. As a result, by and large, names are repeated in every alternate generation. Their religious festivals and ritual celebrations spread over every season of the local calendar.

Most of the ceremonies, feasts, rituals, festivals that are communal in character recur annually and are performed in a particular season, which act as the milestone of the ritual cycle. They worship the above Gods and Goddesses for the welfare of the village people to live peaceful and comfortable life. Witchcraft also plays a prominent role in their lives, which creates suspicion and internal conflicts.

The Goddess *Gangalammatalli*, village deity, is temporarily kept in the centre of the village to protect villagers from outside evil spirits and others. It permanently resides in the forest near the village. *Gangalamma* deity is particularly believed to be associated with good health, abundant harvest and reproduction.

Kodadevatha is the God of the mountain and also the creator of forest, who is believed to be residing in the forest who protects the village people in the forest from outside evil spirits. *Bhumidevatha* is the earth Goddess responsible for yielding good cropping.

The location of their God and Goddess in the forest is covered by the bushes, it is not prominently visible. At the time of celebration they clear the bushes to sacrifice the animals. *Goravadu* who is the *Pujari* of the village maintains the Gods/ Goddesses and celebrates all the festivals, ceremonies related to them once a year. The villagers sing and dance during the celebration of the festival. Earlier they used to sacrifice particular animals for particular Gods/Goddesses as part of the ritual. Due to scarcity of animals it is now replaced by the sacrifice of goats and poultry available in the village.

Besides, in every house, villagers worship family deities. They believe that the direction south is holy and place their deities there and pray atleast once a day.

Konda Reddi rigidly believe that some of the diseases, death and illness of the body is caused by evil spirits and witchcraft. If any person falls ill in a village first a medicine man or magician is called to find the cause of the disease. After knowing the symptoms of the diseases, he provides medicine for healing.

Life Cycle Rituals

The life cycle rituals practiced by the Konda Reddi are (1) childbirth (*Puttuka ritua*), (2) naming ceremony (*Perupettadam*), (3) puberty ceremony (*Peddamanishi avvadam/Samartha or Lavarsala*), (4) marriage ceremony (*Pelli*), (5) pregnancy ceremony (*Pelladithoonddam/Srimantham/kaduputhoundadam*), (6) death rites (*Maranadenalu/Chachhipovadam*).

Childbirth (*Puttuka ritual*): Mother of the pregnant women or any other female relatives who assist her would be rewarded. There is also an elderly woman who performs the function of the mid-wife during every delivery and is appropriately rewarded for her service. Reddi believe that birth of child is due to the blessings of the ancestral spirits. They also believe that the survival of the child is always doubtful unless the ancestral spirits protect the child. At the time of pregnancy and delivery, Reddi follow certain taboos. One such form of taboo is that a new mother of a child is not allowed to cross the water streams near the village. Because the delivery period is treated as a polluted period. During this unclean and polluted period if any woman crosses the streams or any water cannals she would be annoying the Gods and Goddess. It is also believed that the life of the child will be in danger due to the annoyance caused to the Gods and Goddesses. The new mothers are also prohibited from attending any work associated with subsistence, such as going to field, collecting forest produce and fire wood etc. If this taboo is not respected they believe that it will bring a curse to the family and result in low yield of crops. The new mother and child should not come out of the house for one month and also should not go near the *Gondi*, because it will pollute the ancestral spirits which will then lead to scarcity of water, food grains, and also cause damage to the survival of the family members. The last day of the polluted period, that is the day on which the remaining part of the navel is completely separated from the body of the infant, the mother is taken to the stream or any water cannel, where with the assistance of elderly women (*Boddu buddi*), she takes the auspicious bath and buries the umbilical cord and washes all her clothes that were used during the period of seclusion. She also prays to the deity *Gangalamma* to protect her and the infant from all the evil spirits that may haunt them when she comes to fetch water. The spirits are satisfied with the mother and her child and thus many are protected from evil spirits.

Naming Ceremony (*Perupettadam ritual*): A new born baby is given a formal name on this occasion and celebrated with all close relatives by offering every invitee a sumptuous ceremonial meal and liquor. There is no rule as to when a child should be named. In some villages naming ceremony is held after six months and in some after eight days. The function is celebrated with all near and dear ones by having a ceremonial feast, which consists of rich food and the cheeriest local toddy.

Puberty Ceremony (*Peddamanishi avvadam/Samartha or Lavarsala*): Puberty is a significant life cycle occasion in the life of Reddi women. They believe that menstruation is an unclean process and the girl is considered as impure during this period. On attaining puberty, for a week the girl is not allowed to move out of the hut. When the girl attains puberty she takes a normal bath after the 3rd, 5th, or 7th day. Till then she is not allowed to see any men. She is made to sit in a corner of the house on *Tati* (Borassus palm) or *Etha* (palm) *matta* (mat). Till the ceremony is over she has to eat only cooked pulses (*Pappu*) and rice (*Annum*). It is a taboo for any woman or girl to step into the fields or near the Gods and Goddesses during the menstrual period. They believe that if these regulations are not adhered to, the Gods and Goddesses will get annoyed and this will result in low harvest of crops. They are guided by the pollution concept and are allowed to lead a normal life only after the ritualistic bath. *Reddis* believe that puberty heralds the stage of adulthood from childhood and the girls need to be protected from the different evil spirits. To protect from the evil spirits they celebrate a puberty ceremony, which they locally call *Samartha*. The function is celebrated on the 9th or 11th day when she is given a ceremonial bath. On this occasion, first they make the girl to sit on a chair by giving betel leaf, a coconut, a banana in her hands. After this five married women called *parental* start the ceremony by sprinkling *Ackcinthalu* on her and give two bananas into her hand. Later on, all other married women who are invited for the ceremony give their blessings (*Aashervadamu*) to the girl by sprinkling *Ackcinthalu*. The function is celebrated grandly by inviting all the villagers and close relatives. On the occasion of this function they cook *Pulhora*, chicken/ mutton, vegetables, rice and sweet preferably *Jangir/jilebi* etc. All the people who come for the ceremony give gifts in the form of money, utensils, and cloths to the girl. On attaining puberty, the girl is considered to be ready for marriage. They consider that a girl attains complete maturity and she is ready to hold responsibilities and duties of a complete woman.

Marriage Ceremony (*Pelli*): It is a ritual ceremony celebrates in the presence of the village members and close relatives. A detailed description of the marriage is mentioned earlier and the use of natural objects discussed later in the next chapter.

Pregnancy ceremony (*Pelladithoonddam/Srimantham/kaduputhoundadam*): The Konda Reddi celebrates this ceremony for the safe delivery of the pregnant women.

Funeral Rites (*Maranadenalu/Chachhipovadam*): *Konda Reddis* believe in rebirth after death, which they locally called *Puttagatulu*. The dead come back to this world through another birth, so their last rites must be performed in an auspicious way. The spirit of the dead should be satisfied so that they can be reborn. Death associated rituals form the final rites when an individual passes away. They use various natural objects to perform the ceremony. Two to four persons go to the forest to bring the bamboo and prepare the funeral bed for the dead. Once a person is dead all the women in the village give a ritual bath to the dead body and later the relatives dress the dead in new cloths and give some rice to the dead body. Symbolically they use bamboo sticks (*Veduru Karralu*) with which they prepare a layered mat (*Padi*) to carry the corpse to the burial ground. While carrying the corpse in a procession, they throw all varieties of grains mixed in a basket, which is locally called *Akchintalu* on the corpse. They throw these grains in order to satisfy the spirit that is believed to be hovering over the village. Then they bury the dead body by keeping some cooked rice in the hands of the dead along with *Palachettu* leaf. And every one who comes along with the dead body to the burial place also throws some cooked rice along with *Pala* leaf on the dead body. After throwing cooked rice on the dead body they cover the body and all the people go to the stream or bore-well to wash their feet. After this they go to the deceased person's house and sprinkle some *Ackchintalu* in his house and go back to their respective houses and take a bath. On the third day the family members of the dead perform a ritual called *Chinnadina karyam*, which is also called *Chekka chitakadam*. The widow of the dead (only in case of a male deceased) or any elderly widow collects a few plants of *Palachettu*, *Neradu* (rose apple), and *Mamidi* (mango) wood and squash by mixing turmeric and mixes it in a pot full of water. She also cooks some rice and red gram and goes to the burial ground accompanied by the villagers and relatives. She distributes cooked food in three bowls made of *Adda* leaf and all the assembled members take the water from the pot and sprinkle it round the tomb. Reddis mix *Neredu, Mamidi, Pala chettu* and turmeric in water as a purification ritual. Later the family of the deceased offers a feast to the whole village. A year after the death of an individual, his/her family performs a similar ritual called *Peddadinakaryam.* Until this ritual is performed, it is believed that the spirit of the dead wanders but soon after *Peddadinakaryam,* it goes to the *chikatilokam.* Reddis generally bury the dead body of young people and cremate the old people. If a pregnant women dies they strike *Udike manu* nails to the hands and legs. *Reddis* also apply *Chedi* to the eyes and burn the body. They believe that if a pregnant woman is not burnt in the above fashion she may turn into a ghost. Reddis also believe that diseased persons may turn into ghosts if procedures are not followed. Villagers believe that the person who knows Mantras can see ghosts and as can dogs. Reddis also believe in the concept of hell and heaven. A person who does good deeds goes to heaven and otherwise goes to hell.

To sum up, the festivals are usually officiated by the village headmen. All the life cycle rituals and festivals are celebrated under the supervision of the priest (*Pujari/Goravadu*) or *Vejju.* He chants the mantras for appeasing the supernatural forces, asking their blessings for health and happiness. The festivals and rituals are marked off by animal sacrifices to propitiate the Gods and Goddesses, so that they may avert the evils that might befall the community. These are intended to please the Gods and Goddesses, so that he/she shall avert the evils that might befall the community. Reddis consider the religion rites are designed to propitiate the supernatural powers on whose favour they depend. Emotional reaction accompanies the worship of various deities. Their conceptions about the attributes of various deities are vague. They cannot clearly explain the concept of God and Goddess but they certainly believe in God and Goddess, as their forefathers have been doing and carry on the tradition.

These days they worship all Hindu Gods and Goddess and it is evident in many rituals, festivals and ceremonies including life cycle rituals. On the other hand, the spread of Christianity through the different churches and denominations has brought a change in the sacred, social, cultural life of the Reddis in the agency areas. During the fieldwork it has been noticed that there are number of missionaries with different names of churches operating in various scheduled Mandals in east Godavari District. The transformation of the Reddis cultural ceremonies from cohesive group dances, single dances, sacred dances, social dances and group songs to film songs and exhibition of films reflects the gradual changes. Increased contact with non-tribals and more economic dependence on outsiders is leading to an erosion of values and cohesion among the Reddis community.

Food and Drinking Habits

Reddi consume different kinds of cooked foods in a variety of combinations. As a part of their daily diet the villagers eat staple food like rice (Biyam/ rice), *Sode, Sama, Korra, Kandi* etc they eat staple food with pulses like *Srikandulu, Kondakandulu,* (Redgram), *minupappu* (black gram), *pesarpappu* (green gram), *bobbarlu,* and vegetables such as beans, brinjal, tomato, pumpkin, beans, ladies finger, cabbage, cauliflower, in the form of curry. Most of the time rice is consumed either in gruel porridge or boiled form. *Sode* is consumed in morsels. They also consume various types of wild leaves, mushrooms, tubers, roots, fruits and bamboo shoots, etc. that are available in the forests.

The villagers eat the flesh of goat, sheep, chicken, and fowl besides the meat of wild animals and fish. They do not eat beef. Though they do not have any totemic significance, Reddis think that beef and pork are impure meat, which is consumed by other inferior tribes such as *Koya, Kondadora, Kondakammara, Porja and Valmiki,* who are placed lower to the Konda Reddi in the local tribal hierarchy. They take food three times a day either in the form of gruel or morsel. Villagers strictly follow a kind of rationing in order to reduce wastage; they believe that if they waste cooked food or grains, God will curse them in the following harvest season by not yielding them sufficient crops.

The villagers eat seasonal leaves, tubers, roots and fruits in different forms, raw, ripened, and cooked. It is observed that they follow a kind of planning in consuming their harvested grains. All the families store food grains to consume during rainy season since that is the time people are fully involved with agricultural activities and whatever they do would not produce anything immediately. They preserve food grains in order to overcome scarcity of food. The main food they take during this season is *Sode* gruel and preserved dried jackfruit slices in a cooked form. They also eat the pith of *Caryota urens*, mango nuts, *Palmyrah Palm*nuts, and tamarind nuts as food at the time of scarcity.

As most of the households possess 'kitchen gardens' near the house and in the fields they grow different kinds of vegetables and fruits. In the early morning they eat *Salladannam*, which is leftover food from the night mixed with the morning gruel. People believe that this foodstuff provides a lot of energy and strength.

Toddy collection is an important aspect of their life and they consider toddy as food. All villagers consume toddy irrespective of age and sex. Toddy is tapped from palm trees like *Jeelugu Kallu (*Caryota urens toddy) and *Thati Kallu* (Palmyrah toddy), which are abundantly available in their habitat. Every household owns certain number of toddy trees either in their houses, fields or in the forest. Some trees that are found in the forest are also owned by some of the villagers. The *JeeluguKallu* is available from the month of November-December every year and is plentiful during the month of February-June. *ThatiKallu* is available through out the year. Most of the time, the villagers, after the harvest, spend their time in the midst of *Caryota* and *Palmyrah palm* trees consuming the toddy. During the festivals, which fall in the same season, every individual consumes toddy.

Youngsters are always busy with the collection of toddy and the toddy is collected almost three times a day. The toddy collected is brought to *Shavdi/Tatipaki* where interested persons sit together and share the toddy, free of cost, if it is sufficiently available. Some of them sell the toddy and get

Rs. 5/- per *Dokki*, a traditional container made of *Sorakaya*, about 30ml for each fill.

Thus, drinking toddy has ceremonial importance among Reddis. The indulgence of the community in drinking during religious and social ceremonies increases their spirit of festive moods and is expressed during the exhilarating ceremonial dances. During the whole season of its availability, people mostly consume the toddy along with some vegetable or meat in order to reduce the consumption of food grains. During this time, if any stranger, relatives, or any individual in the village visits them, he offered *Jeelugu Kallu* or *Thati Kallu*. Sometimes they also preserve toddy in large containers *(Binde/Donki)* for few days and consume it at the time of agricultural season. People say that unless they consume toddy, they don't feel energetic and cannot work hard in the field.

Some people are also prone to consuming illicit liquor, which is available at shandy places. It is a common scene to come across people returning to their villages from shandy in an intoxicated condition. In addition, they prepare country liquor i.e. *Ippa Sara, Bella Sara, Chiguru* that is made by jaggery.

They prepare *Ippa Sara* by distillation using the dried flowers of *Ippa* tree. People of the village collect Ippa flower during summer and preserved them for winter season to prepare liquor. Liquor contributes to their income as they charge Rs. 10/- for glass. They are also expert in making *Bellam Sara* by using jaggery through the process of distillation. To prepare this, people invest some money on jaggery with an intention of making some money. Earlier they were making this liquor for their own consumption. But these days they are preparing it for making some money. During the off-season, they drink traditional liquor and other modern drinks.

Smoking *beedi* (Akuchutta) and chewing tobacco, eating *Gutaka* are common among both men and women. A strange habit is that some old men and women of the village smoke cigars keeping the lighted end inside their mouth. Both men and women chew the fibrous portion of the tobacco leaf. The habit of smoking cigars and *beedi* is prevalent among children who are above ten years of age. Both parents and younger children together smoke and sometimes share cigars. Smoking and consuming toddy during work is quite common among Reddis.

Blacksmithy and Carpentary

Blacksmiths and carpenters among the Konda Reddi community are part time specialists who work for cash and for agricultural produce. However, there are only two of them in the village. Arjun Rao whose occupation is carpentry, and Ramchander who is a blacksmith. They are professionals and they work for supplementing family income during the season.

Professional blacksmith is generally engaged in making and repairing of various agricultural and farming equipment and gets remuneration, either in cash or in the form of agricultural produce from the owners of the implements. They charge two to three hundred rupees for making plough, sickle, crowbar spade etc. and for repairing, they charge a nominal amount. The carpenter is engaged in making household furniture like door, windows, chairs, tables etc. He works on payment basis for the service. Villagers reported that both of them had come from the plains.

Musical Instruments, Songs and Dance

Reddis are very fond of recreation. Every celebration at family or village level is closely and simultaneously associated with recreation. Folklore includes folk songs, folk dances, folk stories, and other musical instruments. They play music; sing their folk songs and dance at the time of merry making. The villagers name their musical instruments *Dolalu*. The *Dolalu* consist of *Dappu* and *Dola*, a kind of instrument resembling drums. Both the drums look similar but differ in sizes. The *Dola* is larger than *Dappu*. The villagers locally prepare their drums with *Burugu* wood and goat or sheep skin. Both the instruments consist of hollow cylindrical wooden body covered with goats or sheeps' hide on either of the sides. Some kinds of leather strips are tied so as to hold both sides together. They insert small wooden stick to keep the hide firm and stretched to get perfect beat and sound. The beat is rhythmic and loud. At the time of playing both the instruments are hung from the neck of the main musician. While the main musician plays on the right side of the instrument with a stick the second musician plays by using two sticks by following the base drum beat of the first musician. The rest of the drums follow the first drum according to the beat.

Generally the dancers, either in circular or semi-circular fashion follow the musicians by holding one another at the waist and moving according to the rhythm of the drums. In the village, people, irrespective of the age and sex, participate in dancing. The songs are slow, long and with stretching tunes. The songs are tuned to suit the rhythm of the drums. Generally, one individual is the main singer and he/she starts and continues the lyric. The main singer also leads the group of dancers while the others sing the chorus. The folk songs of the *Reddis* reveal their traditions and customs. There are several special songs, which are sung at various special ceremonies and occasions. In their songs they mention the names of their Gods, Goddesses, ancestors, *Podu* fields, agricultural activities, forest, hills, streams, different varieties of flowers, animals and so on. These also reflect their intimate relationship with their environment. They also sing in different lyrics related to the festivals, love and erotic incidents, which have local historical significance. This reveals the attitude and legendry of Reddis towards their natural habitat.

Folk songs, music and dance of the Reddi are a composite whole, which now slowly, reflects the changes. The agricultural ceremonies and other festivals have not changed but during the nights, loudspeakers blare songs before films are screened, and watching movies in the weekends are the observed changes among the Reddi.

Education

In the study village, the older generation could not earlier avail themselves of the opportunity of formal education, because there was no school in the village or close by. Since 1994, the village has an ITDA primary school by the name Girijana Vidya Vikasha Kendram from first standard to class-V. Because of the establishment of the school and the effort of the school teachers, the villagers realized the importance of education and started sending their children to the school in the village. The primary school, which was started in 1994 has two tribal teachers. The village Utlapalem also has one nursery school called (*Anganwadibadi*), which is started recently for the children below first standard. *Anganwadibadi* school has one teacher who distributes food to the school children. The Government provides mid day meal for the school children i.e. 3 Kgs. of rice to each student for a month. After completion of class-V exam, students have to go out of the village for further study to Addateegala, Choudavaram, Yramavaram, Raipalli, and other Mandal head quarters, which is expensive. Very few parents are interested in sending their children for higher studies. The following table gives the education level of the students in the study village.

Table 5
Distribution of Total Educated Villagers at Different Levels

	Village Utlapalem		
Class	*Male*	*Female*	*Total*
I-V	31	22	53
VI-VII	17	06	23
VIII-X	10	08	18
XI-XII	06	04	10
Degree	01	00	01
Total	65	40	105

The total educated population of the village is 105, out of which the male educated population is 65, constituting 61 per cent and female population is 40, constituting 38 per cent. The following table shows the school going children in the village at different levels. The total school going children in the village is 85 out of which the male population is 46, constituting 54 per cent and female population is 34, constituting 40 per cent.

General Sanitation and Health Care Facilities

Villagers are conscious about the health and hygiene though sanitary conditions are poor among Reddis. Generally, all the villagers look healthy and strong. Their regular maintenance of food habits and the quality and quantity of food is naturally nutritious and hygenic. They brush teeth by using *Neem*, and *Parmi* twigs. The youth and middle aged villagers use brush and paste to brush teeth and for bath they use soap. Every individual in the village generally takes bath once in the morning or in the evening after completion of the work. They clean their clothes with the help of a soap or detergent powder.

They sweep their houses everyday and smear the floor at least once in a week. They also white wash and decorate their houses once a year on the occasion of any festival or ceremony. In the study village, all the houses look clean.

There is no drainage system in the village, to drain the domestic sewage water. A number of ditches and pits are seen in the village, which breed mosquitoes and other flies. None of the houses in Utlapalem has latrines and all men, women and children go to their surrounding fields, forests, and ponds etc, for defecation.

As far as diseases are concerned, the general diseases they suffer from, are fever stomach pain, cold, cough, jaundice, malaria etc. They also suffer from fracture, burns for which they use their local medicine, and also go for modern medicine. For the treatment of the diseases they use their own ethno-medicine otherwise they go to government hospitals. If not cured, there are many private dispensaries to visit at the time of need.

The *Reddis* believe mostly in their traditional practice of medicine. Religion plays a vital role in their social life and it influences them in securing health and curing the diseases. The traditional medico-religious practices are done by medicine men (*Vidiyudu*), *Goravadu* and *Vejju* chief medical practitioners, who attempt a kind of ritual called *Neeruparalu* to know the nature of the diseases from the village deity and prescribe some herbal medicine according to the revelation by the deity. The medicine man of the village generally treats the common ailments by using various plants' extract.

They generally use certain herbs and roots in treating various diseases as home remedies. They usually use turmeric and *sonti* (dried ginger), garlic etc for curing, cold cough and fever. They apply herbs and plant latex for treating body pains. But now they are using modern medicine to cure the above minor diseases.

Government Health Care Facilities

A Primary health care (PHC) centre is located at Duppulapalem, which is 10 k.m. away from the village. This centre serves is open all days of the week except on Sundays and other notified National holidays. The PHC opens at 8 a.m. in the morning and closes at 5 p.m. in the evening. The Government hospital located at Addateegala is also made used by the people in an emergency. The doctors, nurses, pharmacist and staff come from other village and those who reside in the same village generally attend to the evening duty.

Multi-Purpose Health Workers (MPHW's) visit the village once a week. They generally work for universal immunization and anti-natal programmes. In addition to this, there are health workers like public health nurse, malaria worker, health assistant who visit the village once a week. The duties of the above workers are to collect the blood sample and distribute medicine like chloraquinine, paracetmol, and other basic drugs for malaria, fever etc. Community based health workers belonging to PHC are employed for distribution of medicine and, there is one-trained woman to per village to help the people with deliveries. A private Registered Medical Practitioners (RMP's) available at Addateegala provide, medication for minor ailments and injuries at a nominal price.

In the village Utlapalem one quack comes thrice a weak. His name is Venkateshwar Rao, he hails from Raipalli village, which is three km from the village. He is a college drop-out Bachelor of Arts. He has done private medical practice in Amalapuram, which is a three-year course. He is trained in

Allopathy and treats diseases like Malaria, Jaundice, Diahoria, Tuberculosis, and Normal fever. According to him, these are major diseases, which are more prevalent among the Reddi. The villagers trust him and consult him for health problems. He recommends the hospital when he feels inadequate to treat the diseases.

Communication and Transport Facilities

Though, the village is located in the interior it is well connected by road with the neighbouring village Anukulapalem which is about 2 kms. The nearest telephone facilities are available at Mandal headquarter Addateegala. For postal facility, and GCC they have to go to D. Ramavaram village, which is 3 kms from the village. Bus services are available thrice a day, connecting to a non-scheduled Mandal called Yeleshwaram. Villagers use bus, jeep, and lorry for traveling to the shandy (weekly market) at Addateegala.

The villagers are availing themselves of some facilities like government school, *Anganwadi* Centre, community hall, sub-branch post office, health Centre, veterinary health centre, VSS centre, Temple, thatched church, 4 hand pumps, youth club, VCD movies, and newspaper etc. The total households of the village are 57, out of which 7 households don't have electricity facility. In the village youth play sports like cricket and volleyball but they do not have any play ground. The wasteland is a playground for them. Because of the above facilities villagers are aware of important happenings outside the village. Due to this close relationship with non-tribals, the Reddi have been influenced socially, materially and culturally.

Voluntary Organizations Working in the Areas

There are some voluntary organizations are working in and around the village Utlapalem, which carry on with many developmental programmes in the village. The European Union Commission development project (STEP) "Sustainable Tribal Empowerment Project" is a process oriented and multi-sectoral initiative. It aims at significant and sustainable improvement in the livelihoods of poor people living in marginalized tribal communities in the agency areas of four districts in northern Andhra Pradesh (Rampachodavaram in East Godavari, Paderu in Visakhapatnam, Parvathipuram in Vizianagaram and Seethapeta in Srikakulam). The project was launched in November 2001 and it is implemented by CARE-India through 29 local NGOs in collaboration with ITDA (under Department of Tribal Development, GoAP) and Community-Based Organizations (CBOs). The project outreach is 2, 35,000 tribal households in 6,200 village/habitations covering 42 mandals spread over 4 districts. All CARE-STEP's partner NGOs are ARISE, CREATOR's PRAGHATHI, SWEEP, SHAKTI and SUJANA.

Other NGOs working in the Mandal headquarters are GIRIJANA DEEPIKA, LAYA, AWARE, ANTHRA, and others. Girijana Deepika was started in 1989, as an independent adivasi peoples' organization to struggle against oppressive forces, restructure the existing inequitable power relations and gain control over their natural resources. The group is composed of Konda Reddi, Koya Dora, and Konda Kammari adivasi groups of East Godavari District. The NGOs provide native medicine to the villagers making them aware of the beneficiary programmes of the government. Encouraging villagers to send their children to the school; protesting against the deforestation that is caused by non-tribals, who are secretly exporting wood from the surroundings of the village. They represent many problems faced by the community to the Mandal and districts headquarters officials and see that the problems are solved. Apart from these, GCC provide cooking provisions for which they are given ration cards and ITDA Rampachowdavaram, is also actively working for the development of the tribals in that area.

Though on the whole, the traditional culture and values of the Konda Reddi has been eroded due to the modern consumerist trends in society, through external forces, the spread of Christianity, etc. there have been positive changes too. Reddis who have been influenced by government programmes and the non-tribals' way of cultivation of lands have accepted the use of improved seeds, fertilizers, and introduction to cash crops. This has vividly affected the self sufficient and sustainable agricultural practices of Reddis in the district. Due to land alienation and lack of land, the Reddis are being forced to migrate. Due to the increase of non-tribal population among the Reddis community there has been a tremendous threat to the cultural identity of the Reddis. Ideals, language, food habits, dress pattern, social, economic and the cultural life of the Konda Reddi have undergone a remarkable change and have been characterized by heterogeneity.

3

Local Knowledge of Agriculture

Agriculture is one of the important economic activities of man. As agriculture fulfils the basic needs of people. Konda Reddis are completely dependent upon natural resources such as land, forest, and animal species available in the region. Land is the primary means of production. The main occupation of the *Reddis* is agriculture. *Reddis* in the study village practice both shifting and settled agriculture. Cultivation, crafts and labour are the main source of livelihood. Those who have plain agricultural lands stopped practicing shifting cultivation. In the present village very few families are practicing shifting cultivation. They cultivate paddy, cereals, millets, vegetables and pulses on the *Podu* slopes and in plain areas. In the village almost all go for labour in the fields, forest, and other works. They also go for hunting, fishing, and collection of minor forest produce (MFPs). They sell some MFPs so as to add to their means of living. The *Reddis* live in their fields in the high and low slopes of their *Podu* and settled agriculture, during the cultivation season from July to January. In the present village nearly 20 families go and live in their fields during the agricultural season and come back to the village after the season.

This chapter discusses Reddis local agricultural knowledge by focusing on the agricultural land classification etc. Attempts to show farmers' production systems and forms of utilization of natural resources among the study community.

Landownership

The total agricultural land owned by the households of study village *Utalapalem* under different types of cultivation. *Reddis* largely depend on dry land cultivation due to non-availability of more wetland in the village. 57 households of Konda Reddi in the village possess 62 acres of dry land, 44 acres of wetland and 12 acres of *Podu* land with a total of 118 acres. However, while 57 of the 78 households or 79.41% possess dry land, which are dependent on rainfall, 44 households or 64.70% have wetlands. Majority of the households in the village own dry land fields. Where as only few families have *Podu* fields in the village. In the village out of 57 households of Konda Reddi 50 households are engaged in land cultivation. On the other hand, 7 households are landless in the study village.

Local Classification of Land Use

Konda Reddi have developed a complex system of agriculture and intensive land management that has survived the test of time and the vagaries of environment. The Reddis have a wide knowledge of the local agricultural land type. Knowledge of land types and their quality and potential for agriculture use is quite high among *Reddis*. Reddis identify and classify lands by soil colour, texture, fertility and taste. Reddis have elaborate knowledge on quality of land type needed for different types of crops to be cultivated. *Reddis* of Utalapalem categorize the cultivable land into four types i.e. *Kondapodu* (dryland). It contains small stones. *Merakabhumi or Garavubhumi* (low dry land) sandy soil; it does not have good moisture holding capacity. *Polambhumi* or *Kuntabhumi* (wet lowland); the land is mostly used for rice cultivation. Soils are often differentiated by colour they are black grey and reddish. They also classify on colour shed i.e. greyness/brightness and darkness/lightness and texture like hardness and smoothness of the soil. *Kondapodu*, land that is located on the hill slopes and terraces, it mostly covered with rocks, bushes and forest growth and the land is used for shifting cultivation. The colour of the soil is light red in colour. *Marakabhumi* or *Garavubhumi* is the one practice on plain jungle clearance and flat land. Sandy soil is characteristic feature of the *Garavu* land. Non-availability of water facility is another characteristics feature of the *Garavubhumi*. And the colour of the soil is brown in colour. *Kuntabhumi* or *Polambhumi* the rainwater or stream water from adjoining forest during rainy season flows into lands and increases the fertility of the soil. This land is more suitable for paddy cultivation, as it needs large quantity of water. The soil is partially black in colour. *Regadibhumi* is black soil land, which is not available in the village and the village farmers practice cultivation in three types of land which are mentioned above. *Garavu* and *Polam* land is used for settled type of agriculture. The following table provides Konda Reddi classification cultivable land.

Individual homesteads, vegetable gardens and *Podu* land come under the category of the individually owned land with hereditary rights of possession. The villagers have the rights of using these lands permanently as long as they used them. But they have no rights of transferring the land to other persons in case they leave the village. In such a situation the village council takes over the possessory rights from the person who leaves the village. The individually owned land includes wetland and permanent cultivable land that are personally owned by individual as their private property. As the economy of the people centers on agriculture, cultivation of land plays an important role in their socio-economic life.

Agricultural Implements

The *Reddis* being a traditional indigenous group still make use of simple but efficient tools and implements that are easily available for various farm operations. These are either obtained from the market or made by them with help of local specialists such as carpenters, blacksmiths, etc.

Almost every family has its own collection of agricultural implements and do not depend very much upon others. The

Table 1
Konda Reddis Local Classification of Cultivable Land

Indigenous land name	*Key characteristics*	*Comments derived from farmers*
Kondapodu	Dry land located on the hill slopes and terraces, it mostly covered with rocks, bushes and forest growth. Sandy and clayey, silty mix reddish in colour.	Very hard to plough, the hardest type of soil. Poor water holding capacity, less fertile, low yielding. Less water consume crops can cultivate. Used for shifting cultivation.
Merakabhumi or garavubhumi	Plain and flat lands dry in nature. Medium textured, when two soil types are mixed together. Sandy loam and clayey greyish brown and very sticky clay.	Mixed with any other soil type, moderate water holding capacity and easy to plough. Soil is Soft when moist but hard when dry. Soil becomes sticky with cultivation. They can cultivate both dry and wet crops but yielding of crop is low.
Polambhumi or *Kuntabhumi*	Low plain land which is wet in nature. Fine textured, sticky soil, sticks to feet while plough. Clayey and sandy greyish brown in colour.	Fertile land mixed with any other soil and better water holding capacity than others, it is easy to plough soil is soft when moist but hard when dry and soil becomes sticky with cultivation. Farmers describe this as the soil use for paddy cultivation.
Regadibhumi	Lowland wet in nature Fine textured, very sticky clay soil, sticks to feet while plough. Clay loam black in colour.	More fertile land very high water holding capacity, very easy to plough, soil is soft when moist but very hard if it does dry out. One of the best soils because it does not require irrigation or fertilizer and it is good for paddy cultivation

number of each type of implements however, varies from family to family according to the land holding size. The larger the land holding, the larger are the number of implements. The number of ploughs to be yoked either to a pair of bullocks possessed by a family is the indication of the size of land holding of the family. At present it is also observed that some families use power levelers or hire for the preparatory tillage of their fields. Appendix 1 shows indigenous agricultural implements used by farmers.

The commonly used implements are plough *(Nagali/ Nanggali)* used for ploughing field, Digging sticks (*Gachala)* or *(Voluva)* used for digging podu fields and used for gathering roots, crowbar *(Gunapam)* is used for digging big area of the field. Sickle *(Kodavali)*, hand-axe *(Goddali)*, big knife or billhook *(Pedda kathi)*, are utilized for cutting purpose. Spade *(Para)* is used for gathering the waste. Large bamboo flat fan *(Chata)* is used as winnowing fan. Wooden forked flail large is used for threshing. Bullock cart is used for carrying the agricultural produce. For storing the grains they use big bamboo containers (*Gadde*). All these are common implements manufactured by themselves with the help of carpenter and black smith. They use different kinds of timbers for manufacturing different implements.

Knowledge of Seasons

Konda Reddi generally perceive three seasons in the year, such as *Varshakalam* (rainy season), *Chalikalam/Sitakalam* (winter season) and *Endakalam* (summer season). And these three seasons are further subdivided into four sub-seasons such as *Shivaranthramasa Kothamavasa Gandhamavasa*, and *Nerelamavasa*. Their seasons start with the onset of rains that is in the last week of May, which falls under *Kothamavasa, which* is closely associated with their agriculture activities. In the *Kothamavasa* season, the people are fully engaged with agricultural activities. It is followed by *Gandhamavasa* and *Nerallamavasa* in which they engage themselves with weeding operations and continue with harvest activities till the end of *Nerallamavasa*. They also carry on hunting activities during this season and also during the later half of *Shivarathramavasa* and simultaneously go in search of new *Podu* fields and initiate *Podu* and other agricultural activities. The seasons are cyclic that is seasonal occurrence takes place in a circular fashion.

Shifting/*Podu* Cultivation

Shifting cultivation is known as *Podu* cultivation in coastal areas of Andhra Pradesh and *Vegad* in *Kolami* and *Padaka* in *Gondi* dialect of Adilabad district in Andhra Pradesh. The *Podu* is the up land and dry land area. The cultivation processes in *Podu* fields start by the end of January. Reddis select that part of the hill side that is well drained by rain water. They clear such land of trees and other vegetation and ready it for cultivation. In selection of the *Podu* land, they rely on their knowledge of local vegetation as an indicator of the fertility of soil. If the patch of land which, grow small size trees and thick bushes it is considered as good land for *podu* cultivation. On the basis of vegetation in a given patch of land, they decide the fertility of land. People also consult the elderly in selecting site for *Podu* cultivation. Easy accessibility of the land, area with out big ditches, rocks, and small trees is preferred. The land should not be under the forest department. Once a man clears a piece of *podu* land it remains in his undisputed possession as long he wishes to raise a crop in it which is generally for three years. As soon as he abandons the field his claim lapses automatically on the land, reverting to jungle, returns to the joint possession of the community. After selection of *Podu* with the consultation of the village *Poojari/Goravdu* the people decide an auspicious day to clear bushes and trees (February to March). Rituals initiate every important phase in the agricultural activities of *Reddis;* coconut-breaking ceremony precedes the felling of the jungle.

A person can clear an average of about one and half acres of forest for *Podu*. For cutting trees they use Axe (*Goddali*), *Chelupukathi* (Billhook). Axes are used for cutting big trees and billhooks are used to clear hard and strong bushes and plants. They also use the sickle *(Kodavali)* for cutting of small bushes. The process of clearing the *Podu* land is called *Tuppacheluputa/*

Podunarakatam in local dialect. They start the process of clearing bushes prior to the onset of summer. They cut the trees leaving one or two feet of stumps on the ground to facilitate re-growth and to prevent soil erosion. If there are big trees, which cannot be cut down, holes are dug around the roots of the tree and a fire is lit at the bottom of the tree. In the months of February and March, trees which are felled dry properly. By the first week of April these trees are burnt down. It takes many days for the fire to spread to all the *Podu* fields. After burning they take half-burnt branches that are reduced to ashes in the fields and the remaining half-burnt branches burnt to ashes in the second year crop. At the time of *Kothamaasa* just prior to the rainy season when the monsoon is about to set in, that is during the month of June, people prepare the land for cultivation. They use the *gachela* or *voluva* (digging stick) for leveling the field.

In the days that follow the Reddi spread ashes and dig the *Podu* land whenever it is necessary. The major task in the *Podu* field is the cutting down of the forest. Women, men, and children take part in clearing and burning the fields. In the family if anybody helps in clearing the forests and wants to cultivate separately, they divide the land into parts, by marking the boundaries with stones, low bunds or half-burnt logs dragged into a line. Sometimes *Hibiscus cannabinus* seeds are sown to mark the boundaries of the *Podu* fields. In these fields the cultivation is taken up for 2-3 years depending on the fertility of the soil. After three years they abandon the site and look for a new site for cultivation. They keep the used *Podu* land fallow till it regains its vegetation and fertility for cultivation.

In the present study village very few people cut *Podu* lands. But still people are practicing *Podu* cultivation on the same patch of the land due to the lack of *Podu* land. During the first week of May, they complete the preparation of fields and wait for the rains. After rainfall they start sowing the seeds. During this time, the *Reddis* construct temporary huts in the field *(chenuillu)*, with Tati (*Borasus palm*) leaves for staying on the field at the time of cultivation. The *Reddis* whose *Podu* fields are far from the village lives in the *Chenuillu* till the harvest.

Table 2
Cycle of Podu Cultivation

Months	*Agriculture activities*
January	Selection of *podu* land
February	Cutting and drying of felled trees
March	Gathering felled trees and bushes
April	Burning of felled trees and bushes
May	Spreading of ashes and digging of *podu*
June	Broadcasting of seeds
July	Weeding operations
August	Watching of crops
September	Harvesting of maize
October	Harvesting of *sama, korra, bontalu*
November	Harvesting of jowar
December	Harvesting of *ragi*

And the fields, which are near to the village, they construct a shelter called *Chenupaka* (field house).

In the second week of June they start broadcasting the seeds called *Tholakarijallu.* Fencing of the fields is done after sowing the seeds *(Pickelu/Vithanalu).* They fence the fields with bamboo and bushes. The sowing of seeds starts with a great ritual called *Bhumi Pandaga* (Mother earth Festival) or *Vithanala Pandaga* (seed festival). On this festival, the *Pujari* (*Gorvadu*) sacrifices chicken to *Bhumi Devata* at the center of the fields and the blood should drip on earth and on the seeds, which are collected from all the households. Reddis performs the ritual of offering a chicken and coconut to the goddess. While chanting mantras Reddis asks goddess to do things in their favour. The *Goravadu* usually conducts the rites on his fields or in centered place in a junction of several fields. The blood of the sacrificial animal is allowed to first spice on the earth and then a piece of it is taken and mixed with the seeds, which they want to sow in the field.

After the ceremony of *Bhumi pandaga,* they go for hunting continuously for three days and sowing of the seeds start on fourth day. After this main ritual each households sacrifices a chicken on their own. They pluck a feather of the sacrificed chicken and stick to the sowing basket. So that the crop cultivation should grow thick like poultry feather. *Podu* land is used for cultivation of crops like *Sama, Korra. Sollu, Mokka Jonna, Jonna, Bonthalu, Nuvvi, Vanga*. *Reddis* said that the early crops (*Jonna Pyru*) like *Mokka Jonna* variety, and late crops (*Enda Pyru*) such as *Korra, Samalu, Bonthalu, Nattupogaku* are not only free from pest attack, but also do not require much water, except the rainwater. They cultivate pulses like *Minumulu, Babarlu,* and *Kandi,* in this land in the season of Kharif. They practice two methods for sowing in *podu*; one is broadcasting of seed over the surface of land and the other is dibbling of the seed into holes made with a digging stick. Broadcasting is for small millet like *Sama, Korra, Sollu, Bonthalu, Nuvvi, Vanga, Minumulu, Babarlu,* According to them, the seeds, which are small in size, required in large quantities are better broadcasted. This also reduces the chance of failure in seed germination. Broadcasting seeds does not need much water, except the rainwater. Whereas other crops like *Jonna, Kandi, Mokka Jonna*, are dibbled into the ground. They put 2 seeds into the holes, so that if one does not germinate, the other might germinate which reduces the chance of failure in seed germination. They sow *Goggu* (*hibiscus Cannabinus*) in between grains and pulses for dividing the crops. The leaves and seeds of *Goggu* tree are used for eating and the stem is used for fiber. Sometimes they cultivate mixed crops of jowar and small millet. Small millet is sown first by men only and seeds of mixed broadcast varieties like jowar are scattered at each step with the left hand. Seeds in a basket called *Parak butta* are hung on the left shoulder. Women carry seeds in a cloth and wrap it around their waist at the time of sowing.

In *Podu* cultivation the practice of mixed cropping is common. There is no scope for monocropping, as different crops are sown in a single patch of land. A large variety of crops such as *Gantelu* (bajra), *Jonna, Bonthalujonna* (jowar) *Gaggerajonna, Kandi* (red gram), *Arlu, Pesara, Babbara, Buduma, Uluvalu, Sama* (small millets), *Nuvulu, Babarllu, Minimulu, Pathi* (cotton), *Pogaaku, Kallipogaaku, Gogu, Sodi* (*ragi*), *Korralu* (Italian millet), *Amudalu* (castor) are grown in *Podu* lands. In *Podu* cultivation the primary emphasis is on food crops, which are meant for household consumption. According to the people the mixed cropping pattern has certain advantages such as, a supply of variety of food grains is ensured, mixed cropping ensures food because there is no problem such as a total failure of crop. It also ensures that at least one or the other crop survives and produce yield.

When mixed cropping is employed, *Kandulu* (red gram) are intermixed with that of cotton, jowar, chilly, *Zea maize*, etc. The holes are made slightly in front of the sower, so that digging easily pierces the earth with his long handled *Palugu* i.e. digging stick and bending forward, drops a seed into the hole. It is not sown in rows but haphazardly. Though the sowing takes place together for all seeds the harvest takes place at different intervals of time. *Sode* (ragi) seeds are broadcasted in the same field, when the previously sown seeds germinate and grow for fifteen days. This is because the harvest of *Sode* grains takes place at the end of the season, since *Sode* grains are very small in size and that needs extra care and concentration at the time of harvest. If the patch is small they broadcast one or two varieties of seeds and sometimes depending on the availability of seeds and also the grains needed for consumption. After the broadcast of above seeds in the *Kondapodu* fields, another type of sowing that is dibbling takes place. Generally dibbling of seeds takes place when the already sown seeds germinate and grow to considerable extent. This takes place in the month of August. They dibble with the *Voluva* or *Gachala* in the gaps that are located between the crops in the field and also at the places where broadcasted seeds do not germinate *Amudalu*, *Alusandalu* and *Anumalu*, *Jonna*, *Kandulu*. They sow single variety of seeds in the holes except in the case of *Kandulu* and *Amudalu* in order to make it convenient for the *Anumalu* creeper to creep on the *Kandi* plant.

Some farmers also use intercropping. The majority of the informants said that the mixing of crops is not common among Reddis. Because, one of the mixed crops can dominate the other and mature faster than it. However, mixed cropping has several advantages. Farmers are well aware that soil with a good infiltration capacity reduces the runoff and consequently the erosion hazard. They try to avoid harmful water logging by constructing drainage ditches within cultivated fields that are useful in drainage of excess water from soils and in providing a good oxygen supply and micro-organism for the roots. In the recent past, they used to plough the land in August and suspend it for some time. They turned it over in October. Again they ploughed it in May, and sowed crops in June. Reddis consider this the best measure to improve the fertility of the land.

Reddis farmers have their own unique varieties of major crops such as paddy, *Gantelu, Jonna,* and *Kandi* etc. These varieties have been developed and maintained over time, in order to cope with a climate that is characterized by erratic rainfall. The farmers have developed a set of strategies to select the seeds, maintain a seed stock and anticipate climate changes. These strategies, which are holistic in nature and include both physical and spiritual indicators, are widely practiced and are of significant relevance to agricultural development.

Dryland Cultivation/Garavu Vevasaayam

When the monsoon is about to set in they clear small bushes and grass and prepare the land for cultivation. The ploughing of dryland takes place during the first or second weeks of June and the field is kept ploughed till the arrival of full monsoon. Immediately after 3 or 4 heavy rains the land is ploughed again. While ploughing the second time the seeds are broadcasted and covered in the sand. Preferably the broadcasting of seeds in the dry cultivation fields take place on a rainy day because the seeds are sown in the moisture. Ploughed field, which results in the perfect germination of seeds. The broadcasting of seeds takes place in the month of July. Mixed as well as mono cropping is prevalent. Generally the grains, which are sown, or not sown in the *Podu* cultivation, are preferred and also which are needed in large quantity for consumption. Though an intensive weeding is not observed but a nominal weeding is enough because the rate of germination of the sown seeds is moderate. The above land is used for cultivation of crops of *Sodi, Kandulu* (Redgram/*Cajanus cajan*), *Babbarlu,*

Minimulu, Goggu, Jonna (Jowar/*Sorghum vulgare*), *Mokkajona* (*Zea maize*), *Pathi* (cotton), *Nvvullu* (Ceasmum), *Gantelu, Chilies* (*Capsicum anum*), *Vanga* (*Solanum melongina*), *Tamata* (*Lycopersicum esculantem*), *Veruchanagalu, Pesaru* (*Phaselos mungo*), *Pogaaku, Gidimamidi* (*Anacardium accidentalis,*) and *Saggu dumpa* etc. *Sodi, Pesaru, Nuvullu* are broadcast in the dry fields where as *Amudalu, Alusandulu, Jonna, Mokkajona, Kandulu* are dibbled and *Pathi, Pogaaku, Saggu dumpa* are sown in the field. Both single and mixed cropping pattern is prevalent in the dry land cultivation. Apart from the cereals and millets, in cool weather some vegetables such as brinjal (*Vanga*), tomato (*Tamata*), ash gourd (*Burda gummadi*), *Dippa, donki* angular gourd, beans, bottle gourd, onions, pumpkin, *thotakura*, and maize etc., are also grown by them in kitchen garden (*Doddi*).

Apart from the grain crops, some vegetables such as brinjals (*Vanga*), sweet potato, tomato, angular guard, beans, bottle guard, onions, pumpkin, etc. are also raised. In the dry land cultivation the harvest starts with the maize in the month of September followed by the harvest of jowar in the month of November and *Sodi* (ragi) in the month of December and finally harvest ends with *Kandi* (red gram), *Adusulu, Anumulu, Alusandulu* in the month of Janurary.

On the *Podu* and plains lands of the *Reddis* in the study area they have planted cashew nut trees provided by I.T.D.A. And on the plain lands they adopted a new crop *Dumpa* (tubers) that they had learnt from their contact with non-tribal *Turpu Kapu* of the village *MulleruMallavaram*. Now almost all of them cultivate this crop because they get more income and also this land is very suitable for *Dumpa* crop. In this crop there are some varieties like *Karra pendlum* and *Faram Dumpa, which* are old varieties with which they prepare curry and also eat them by roasting. These varieties of crops grow in about four months. New varieties in these crops are *Tella Gulabi, Yerra Gulabi, Kerala Dumpa.* These varieties are pest resistant and only manure is enough to cultivate these crops. Reasons for the *Reddis* shifting to these tubers is that the cash crops they cultivate on the planes like cotton and brinjal were not beneficial due to the crop failure and sometimes due to lack of water. The new tuber crop needs less water to grow and the income is good. New varieties of crops grown by the Reddis are mentioned below

Tella Gulabi (crop): The stem of this variety is pure green in colour with big leaves and branches. The tuber of crop is long and big. The stem has big nodes and internodes measuring 24 inches. Each plant has 8-15 tubers. The rind of the tuber is white in colour and the inner part is red in colour.

Yerra Gulabi (crop): The stem of these plants is slightly red in colour at the nodes and the inter-nodal distance is six-inch. The tubers are long and thin in structure. Each plant has 8-20 tubers. The peel of the tuber is red in colour and the inner part is pure white in colour.

Kerala Dumpa (crop): The stem of the plant is long and stumpy and the distance between nodes is 24 inches. This variety needs optimum temperature to grow more and to produce more tubers. This variety takes seven months to give the yield and it is more economically fetching as it produces about 8-30 tubers per a plant. The rind of the tuber is grey in colour and the inner part is white. The tubers are short and fat.

Faram Dumpa (tuber crop): The stem of the plant is pure white in colour and the nodes are very close. The peel of the tuber is white in colour and inner part is white. Each plant produces 8-20 tubers.

All the above-mentioned crops do not need much water and also pesticides because the crops are pest resistant. The fruits of these tuber crops and stems are very dangerous for domestic animals, and if they eat more it can be fatal. By chance if any animal eats these crop *Konda Reddis* do not allow the animals to drink water for five hours. It is said that if the animal consumes water after eating these plants it may die with in three hours. The wild animals like Wild boars, Porcupines etc, which feed on tubers, damage these crops. Appendix 2 provides different varieties of crops grown by the *Reddis* in their fields.

Wetland Cultivation

Paddy is cultivated in wetlands called *Polam* and the fields are enclosed by bunds *(Gattlu)*, where the rainwater is stored. The wetlands where the Reddis practice wet cultivation are in the plains area of the foothills. Such plain land is divided into a number of rectangular or square shaped plots, the boundary of which is demarcated by mud ridges. They have adopted certain innovations of scientific agriculture to some extent, but are still traditional cultivators, for they still use traditional farm practices and techniques that are handed down from the past generations. Mono cropping is the rule in the case of paddy fields. Their paddy fields remain barren for at least six months in a year. In the wetland (*Polambhumi* or *Kuntabhoomi*) ploughing starts prior to the onset of monsoon. Monsoon showers locally called *Dukki Varsham*. After this rain, soil becomes moist and is easily loosened and the people, who possess both *Podu* and plain land, first start ploughing on *Podu*. Men exclusively do ploughing and they go for ploughing early in the morning at 6 a.m. The purpose of ploughing is for destruction of the weeds and leveling the surface.

During summer ridges are built in the field to stagnant rainwater. Ploughing and the clearing of the grass and bushes in the field follow this. From the past they used to manure (*Peda*) their paddy fields with the locally available manure. The excreta of domestic animals mixed with the daily household sweep and other waste materials are stored in pits on the periphery of the habitation of every household at least for six months so that the materials may rot completely. Then the manure is spread over or mixed with soil before the onset of the first monsoon rain to make the fields fertile. At the same time large pieces of land are thrashed and the field is prepared for cultivation. At present, in the village some farmers use chemical fertilizers to manure their fields. Though they responded well to this innovation, they cannot make full use of it because of the insufficient knowledge of its application, lack of ready cash for purchase and uncertainty involved in its availability. They have to purchase fertilizers from out side their villages from the distant places or some times moneylenders provide credit them if they are in need.

During the onset of monsoon the people again plough the field and practice "*Bendakottadamu*" at the same prepare an exclusive patch of land for nursing the seeds by filling a particular patch of land with water and broadcast the paddy seeds, which is called *"Akupoyadamu"* and the patch is called *Akumadi* in the local dialect. Cultivation of mono crop that is paddy is the characteristic feature of *Polambhumi* or *Kuntabhoomi* and it is strictly observed. The cultivation of paddy crops of *Reddis* is mostly rain fed, and some water is used from *Kaluva* and *Uta Neeru*. Water harvesting has done by one field to another fields through bunds. The table presents an overview of various rice varieties crop grown by the villagers. A number of observations on the yield-stabilizing functions of indigenous varietal knowledge and the polyvarietal planting of crops were made in the course of a study of the knowledge and use of rice varieties among *Konda Reddis*. The Appendix 3 provides an overview of various local rice varieties cultivated by the *Reddis*.

The nursing of seeds takes place during the second and third weeks of June that is at the initial stage of monsoon. Some people transplant the seedlings of the paddy crop raised in nurseries in the ploughed and puddle fields. In *Akumadi* where nursing is done, the supply of water is undertaken at regular intervals of time. Mostly rainwater is relied upon but if needed natives carry water from stream near by the field. The germinated plants allowed growing for three weeks. During the second week, weeding is observed. The transplantation takes place after three weeks of plant growth on a rainy day. For transplantation labour assistance is sought from the villagers because the transplantation should be completed within a day the transplantation is known as *" Udpadamu"* in local term. It takes 5-6 months to harvest from the day of transplantation. Mean while the natives weed the field at two regular intervals of time and water the fields according to the requirement.

Once the rain falls, the prepared fields are ready for paddy plantation. At first a small patch of land is prepared and paddy is sown in it to germinate. After fifteen days when the saplings are ready for transplantation, they are transplanted in the entire area. For the transplantation the family members and close relatives provide labour on the basis of reciprocity. Once the transplants are fifteen days old, the entire region becomes green, since the transplanted paddy has attained 20-25 cm of height. After this weeding work is carried out. In this, family also collectively works and removes the weeds. Women engage in weeding. The women of family and lineage join in the process and complete it.

After the crop has full grown in the month of December to January, it is harvested and brought to the *Khallam* (thrashing ground). Threshing is locally called as *Noorpidi*. The place where they thresh paddy is called *Kallam*. For preparation of the threshing place first they clean a place near their fields and then cow dung is smeared. After a day, when the place is dried, they start threshing. At the time of threshing first put some quantity of grass on the threshing place and the bullocks are yoked to a round beam, which is driven through the grains. The *Reddis* make use of *Khallam* for thrashing of the harvested crops. *Khallam* is a large space cleaned near field and is used for thrashing the paddy and other crops to extract the grains. Each household has their own *Khallam*. To prepare this space close relatives and friends provide labour on the basis of reciprocity. Some times families have *Khallam* on the basis of lineage and joint family, since the division of house might have taken place and the agriculture fields not been divided. Therefore they collectively harvest the crop and distribute the product.

After entire crops are brought from the field to the *Khallam*, then with help of oxen or wooden stands, the thrashing of the paddy is carried out. Two to four pair of oxen is tied to the wooden pen fixed in the centre of the *Khallam*. They walk on the paddy spread on the ground. In this process the paddy breaks of from straws, and it is separated from the straw by blowing process using the wind. Pulses and *Nuvulu* is extracted by beating it on a wooden platform. When the entire grains are extracted, two fists full of grain tied in a cloth is tied to the wooden pen in the centre. Other gains like *Pesaru, Minumulu, Babbarlu* are often thrashed with a stick. The threshed gain is dropped from a height, usually above the head with winnowing fan (*Chata)* in front of a winnowing fan or against strong winds, so that wind carries off the husk. Then the whole produce is kept in the store, and ritual is offered to the *Bhudevi* in the *Khallam*. A coconut is offered to her. Thereafter the grain tied to the central pen is removed and used in the rituals kept in a safe place and at the time of *Gangalamma* in the month of May.

The harvest of paddy takes place when the sheaf of corn is ripe. Paddy crop is guarded from the wild animals and pests from the top of an erect *Mancha*. The corn is guarded till it is dry and completely rid of moisture. On the day of harvest *Reddis* cut the paddy crop and keep it on the *Mancha* and a few days later thrash the paddy with the help of cattle and winnow the chaff, husk and grass from grains. They measure the grains with *Kuncham* and carry home and preserve them in the bamboo baskets and pots called *Gudagampa* and *Gudakunda* respectively.

At the time of harvest every household sacrifices chicken to *Bhumi Devata* in their fields. The harvested grain is brought home in bags (*Bastha*), *Buttalu* (wattle grain bin), *tattalu* (baskets) on the bullock carts. The paddy crop, which is brought from the field, is stored in *Gadde*, which is close to the house. Bamboos are split, and *Gadde* is prepared and then it is plastered with mud and dung. The produce is collected in the storehouse and after offering the new grains to the village deity *Gangalamma*, is brought into households for eating. The weeds are used as fodder for the cattle. The season ends with harvesting and thrashing of the paddy.

The plain lands are usually cultivated in partnership by two or more families, who share the yield after the harvest. In this type of cultivation there is mutual cooperation, as some of them possess bullocks and some of them do not. But they all work according to their individual capacity on their common land pooling their resources to pay the land revenue and then share the crops equally.

Besides the crops cultivation *Reddis* maintain home or kitchen gardens as an indigenous method of fertility management on a small scale. Reddis manage a simple and small farm in homesteads, where various vegetables, and fruits like, chili, maize, beans, Cucumbers, *Budama, Dippa, Dhonki, Burda Gummadi* etc., can be grown in kitchen garden named as *Doddi*. Waste drainage water from the kitchen can be used to start seedlings. Ashes and sweepings from the household and domestic animals' manure are normally spread in the garden. The wastewater, ash, and also the animal bedding fertile the soil around the homestead. It is not surprising that a kitchen garden, though small in total area, is an important part of the overall production system and has successfully been used as an entry point into the enhanced productivity of the farming system. The Appendix 4 provides vegetables and fruits grown by the Reddis in their kitchen garden.

In the village *Utalapalem* most people have vegetable kitchen gardens, where they grow vegetables and have planted some fruit trees where the size of the garden is very small in size below half acre. These vegetables are used for household consumption and also some of them sold in the market. Generally they grow in rainy and winter season. The vegetables, which are grown in garden, are sufficient for household consumption.

Beans and other creeper varieties are grown by giving them support with thorny bushes (*Kampa*) or some branches. All these vegetables supplement their diet. Some of them also cultivate these vegetables on a bigger scale in their fields and sell them. Brinjals are sold for Rs. 50-60 per bag and sometime when there is demand in the market they sell a bag for Rs 150-200. Many other plants and fruit bearing trees grow near their houses and fields like Mango, Tamarind, Orange, guava, Jack fruit, lemon. All these become the individual property of the man who planted them and these are inherited by his kin.

Weed Management Practice in Cultivation

Weeding takes place two times in *Kondapodu* during the cultivation period. The first weeding takes place when the *Sama* crop grows to a considerable level that is fifteen days after its germination. If weeding does not take place properly the crops rot. The increase of unnecessary plants in the *Podu* damages the crop unless the weeding is done. Another round of weeding take place in the month of October, much prior to the harvest, which enables the crops to ensure good yield, and also for convenient harvesting. After the weeding process, every farmer keeps a careful watch on the crops in order to protect them from the pests and wild animals. In general they erect a Pandal, which is called *Mancha* in the midst of the field and safeguard the crop by throwing stones at wild animals that destroy the crop and also by making loud sounds by beating tin and simultaneously shouting in order to drive away the animals from the fields.

The harvest starts when the *Sama, Korra* ripens and immediately the village council meet to celebrate the festival *Korrakotha Panduga*, which is a harvest festival. Until the festival is celebrated it is considered as taboo to harvest *Sama* or *Korra* or any other yield from the *Podu*. Among the crops the first ones to be harvested are *Sama, Korra* and *Nuvvulu* in the month of September. From *Nuvvulu*, they extract oil used for cooking purposes by the *Reddis*. The harvest of *Gante, Jonna, Mokkajonna, Bontha, Gijjerajonna* takes place one after the other from the middle of October to the end of November. They maintain a week to ten days gap in harvesting each grain. The harvest of *Sodi* takes place in the month of December. In the month of January prior to the festival of *Sankranthi* the rest of the dibbled pulses crops such as *Kandi, Alusandalu, Anumulu,* and *Amudalu* are harvested. They celebrate *Pappu Pandaga* before the harvesting of pulses. The process of harvest takes place in a chain of stages that is first they cut the sheaf of corn and dry them by keeping it on the *mancha*. And after a few days thrashing is done.

If the yield obtained is high they thrash with the help of cattle and if the yield is low the family members including some hired labour thrash the crop. Where the process of thrashing is done to separate the grain from husk and chaff with the help of winnowing fan, which is called *"Polikattadam"* in local term; after this, the grains are measured and carried home for preservation. The winnowing of the grains is carried out with an implement called *Cheta* and measuring is done using *Kuncham* a local measure. The grains are preserved in different sizes of baskets made of bamboo reeds. These days, *Reddis* are planting cashew trees on *Podu* fields. The *Reddis* are being benefited from this cash crop. The farmers of the village only take single crop i.e. in Kharif season and after that fields are left fallow and it becomes a common grazing land for the cattle of the area. Generally after agriculture season the villagers have no other employment activities except going to the forest for other works like hunting, collection of MFP's etc.

Source and Methods of Irrigation

The villagers mainly depend on the rain for cultivation. There is no alternative source of water except rainy season. Canals are also an important source of water storage for cultivation. Village farmers use canal and stream water (*Utaneru)* for irrigation of the rice crop. They dig some water channels (*Tummulu)* and divert the water from the main canal into their fields. These channels are dug out and vertically cut trunks of toddy palms (*Tati*) are laid as flooring, and the banks (*Gattlu)* are constructed with mud and stones. Konda Reddies use the canal water only when needed and otherwise they close the banks (*Gattlu)* of the cannels. Most of the farmers are learning to conserve the water by digging a pit near the fields and store the *Uta* (stream water) and use. This pit is called as *Pedda Madi* from this *Madi* they supply the water to the crops.

Farmers of the village do not use any engines for bringing water through canals and it is not possible because they lack the money. If there is storage water in the canal they distribute it equally among the farmers.

The villagers mainly depend on the rain season for cultivation. Canals are also an important source of water for cultivation, when droughts occurr in the region. At the time of drought no crops are grown except *Gante* because this crop does not need water. And also buy the *Nukalu* from the shops for rupees four per *Kuncham*. And some of the other crops like *Gante, Sama, Korra, Bonthalu, Allu, and Kandi* are aquired as payment for labour.

The village Utalapalem has canal towards south side, where water gets accumulated in rainy season and it continues till the winter season. Twenty-five farmers use this canal to their rice crop. And this water cultivates around 40 acres of land. For this *Reddis* built a wall in of one its bends parallel to the canal bank in order to divert some of the soil and water from the canal to make the lands fertile.

Methods of Seed Selection and Preservation

In the village Utalapalem the Reddi farmers produce most of their own seeds, and sometimes obtain new varieties from local markets or from other farmers or other sources. Once the seeds are obtained, they are kept for a long period. Various methods of seed selection and preservation are practiced by the farmers of the *Utalapalem* for the food crops and cash crops like rice, pulses, vegetables and tubers. They select the best seeds, with different techniques for different crops. A description of the traditional methods of preserving the grains is given below:

For selecting paddy seeds for the next sowing, farmers select seed during the harvest. The paddy is usually stacked in bundles; and they identify bundles bearing panicles with big, heavy, undamaged and fully ripe grains to be kept as seed. If they select slightly damaged grain, it cannot be stored and will spoil. Farmers also choose panicles with best grain and will cut them from the standing crop to obtain seed by threshing. They do not allow intermixture of paddy varieties and the seeds are stored separately.

The selected grains are first cleaned and dried in the sun. After they are dried well, they are preserved for next year for sowing. Most farmers keep the paddy grains in a gunny bag (bags made by jute Fiber) or *Buttalu, Pindi sanchulu,* plastic polythene or bags. Farmers often store the grains in a special container locally called as *Gadde*. It is made of bamboo or *Juvvi* sticks and plastered by red soil. Both the inner and outer walls of the *Gadde* plastered with mud and dung. Occasionally the grains are exposed to sunlight for drying. This practice is quite useful in maintaining the quality of the seed. Drying is usually repeated two or three times at an interval of 3-4 months depending upon the weather conditions.

Post harvest pests are not a problem in the *Reddis* regions. Most of the produce is gradually consumed by the *Reddis* after harvest. Processed foods are adequately stored in *Gadde* outside smoke filled kitchen.

For generations this practice has been employed by farmers in the village to protect their crops. The bins, baskets or cans used for storing the grains are generally made of thatch bamboo, wood or thin sheets of aluminum. Grain is also stored in large earthenware pots, and many of the farmers in the low-income group make use of bamboo baskets, traditionally *Gadde*, which is made of bamboo and red mud, or gunnysacks (bags made of jute fibres).

The quantity of grains stored in a bag or basket generally varies from 50-100 kg. For every bag of grains neem leaves (*Vyapa Aku*) or *Bodataram*/kukkapogaaku kada (*sphaeranthus indicus*) are mixed together with a few of the more tender branches. The tree branches and leaves are poisonous to the insects. The amount of leaves is increased with the quantity of grains. It has been reported by the *Reddis* that the grains stored together with above leaves are not affected by pests of any kind. If the baskets are kept completely airtight, the occurrence of pests in the grains is reduced virtually to nil. It is also reported that after few months various pests are to be found in the bamboo baskets and in gunnysacks where grains are stored. To prevent this they dry grains every 5-6 months and add leaves. Most farmers have now adopted this practice, which is available at virtually no cost, as the above leaves are in plentiful supply throughout the year.

Pulses

Big and fully ripe pods of the pulse variety are handpicked in the fields for further processing at the time of harvesting the crop. Another common practice is selecting a few healthy plants bearing bold and ripe pods and harvesting them separately. Once the pods have been selected, they are dried and the seeds are extracted. These are carefully dried in the sun before being stored. Pulses are usually stored in *pota, dippa;* mixed with wood ash. After placing the seeds in the pot, its opening is firmly closed to make it airtight by plastering with mud and animal dung. Some farmers' deliberately do not fill the container completely. Storing grains in this way is believed to protect them from pest and disease. However, occasional checks are made, and if there is any sign of damage, the grains are again exposed to the sun, carefully dried, and mixed with fresh wood ash before they are replaced in the container.

Vegetables

When selecting the seeds of fruits and vegetables like bitter gourd (*kakara kaya*), ridge bottle gourd (*birakaya*), pumpkin (*anapakaya*), cucumber (*dosakaya*), tomato, chilies etc., the fruits are allowed to mature, ripen and dry while still on the plants. These matured fruits are picked and further dried by repeated exposure to the sunlight. Then the seeds are extracted from the fruits and are dried in the sun. Bottle gourd, pumpkin and ridge gourd, lady finger *Benda Kaya, alisanda,* chillies (*Mirapa),* beans (*chikkudu)* are kept hanging on the roof for months after being picked. Then they are cut open and the seeds are extracted, cleaned and thoroughly dried in the sun before final storage. In the case of brinjal, farmers sometimes pick the ripe fruit store it in polythene bags and allow it to dry. It is then washed in water and seeds are strained out, then they are cleaned and dried thoroughly in the sun. Sometimes rotten brinjals are cut on four sides and pressed to remove the seeds. Once the seeds have been extracted and cleaned they are thoroughly dried. These seeds are stored in polythene covers. The pulp of the ripened tomatoes along with seeds is removed and it is smeared on the walls. The seeds are taken out once the pulp is dried up on the wall. To ensure the keeping quality of the seeds and to protect them from fungus or pests during storage, they are occasionally taken out and dried in the sun. The seeds are usually dried 2-3 times at intervals of 3-4 months. The Maize cobs along with the seeds are wrapped up with a thick layer of leaves of the same plant and are stored.

Pest Management

Reddis consider birds, domestic animals and wild animals as pests to the crops. Because they all damage crops and vegetable fruits. When such problems arise such as attacks by all the above animals and birds, locally available materials and knowledge were used to reduce losses. These traditional pest control practices formed an important component of Reddis farming systems, contributing to their flexibility, productivity and sustainability.

The major pest problems faced by Reddis in the region are wild animals like wild pigs *(Adavi Pandi)*, porcupine *(Mullapandi)*, and *Pandikokku, Murdaeluka* which spoil crops like *Dumpalu* (stem tubers), Cotton crops are spoiled by animals like *Muradaeluka, Ellitaeluka*, Cows, *Gorragedde, Gorre, Kothulu* (monkeys) and *Bandalu*. And pulses are spoiled by rabbit *(Chevilapilli), Gorre, Kanasu, Gorra Gedde, Kotulu*, Maize crops are spoiled by birds like parrot (*Chelaks)*, crow *(Kaki*), squirrel *(Udutha Eluka*), cows, *Gorra Gedde* etc.

To prevent rats and birds from gaining access to maize some farmers protect developing maize cob by enclosing them in a large breadfruit leaf occasionally. The wrapping is done with maize leaves itself. Reddis believe that the rats and birds do not like to eat the maize cover. Rats and birds are caught by means of stone or stick traps that are made using a large flat piece of coral stone levered over bait of dried fish or grains like *Gantelu* or mango seeds. The trap is triggered using a tripwire made from the fibre and woods or bamboo sticks. Sticks traps are similarly constructed, using wood instead of stones, which *Reddis* locally call *Adapa*. For birds, *Uchus are* fixed in the tops of the trees and for animals they are fixed at the corner or middle of the fields during the nights. For protection from big animals like wild Pigs, *Mulla Pandi, Duppi, Pandi Kokku, Murda* they construct *Kuddelu* at the corners and side fields to trap the animals. Kuddelus are constructed with the help of heavy stones. If an animal enter that trap all stones fall on the animal and ultimately it will die. Vertebrates cause the major pest problems in the fields. To combat attacks by major crop pests, a variety of resourceful traditional pest control practices are employed, such as those described above. The problems caused by insects and plant diseases have tended to reduce by spreading neem fruits powder. Although a number of these practices, which are described above, are still used today for vertebrate pest control, most farmers confronted with new pest problems have opted for introduced technology.

And in the fields almost all crops protected by hanging cloth or polythene bags etc. are placed on the trees and on the crops, where they are locally called *Bedaddlu*. And also to protect the crop *Reddis* tie *Pala Teega, Veduru Poya,* and Iron wire around the field so that ward off the animals.

With the contact of non-tribals Reddis have started using the maximum amount of fertilizer. The principal chemical fertilizers in use are monocrotopus, naucron, endosulphon, Dithane M45, Dithane7, Di Amonium Phosphate, Super phosphate, and Urea etc.

Soil Erosion and Local Soil Management Techniques

According to the *Reddi* farmers, the causes of soil erosion involve highly erosive rainstorms, overgrazing, deforestation, population growth, wind, traditional farming system and the acts of God. When the land becomes full of pits, the rain erodes it and the soils are broken up into fine particles, they will be vulnerable to erosion. The preparation of the land for rice (Vari), and other small seed crops accelerate soil erosion. They cleared forests for agriculture and pastures. Deforestation resulted in the degradation of soil resources, once the nutrient flush from conversion of forest to pasture is over.

Reddis have employed different conservation practices to overcome the problems of environmental degradation. These practices include crop rotations, direct additions of nutrients in the form of application of ashes, mulches, residues, cattle dung, composting, contour ploughing, fallowing, contour ridging and furrows, terracing, arable-pasture rotation, planting trees, and so forth. *Reddis* have used cattle dung as the best fertilizer. They know the different values of the excrement of domestic animals. The first important dung is cattle dung of cow, ox and buffalo is believed to be superior to other forms of dung in highlands. Most of the informants in the study area agree that the excreta of sheep and goat is also important, for they penetrate the land and enrich the soil. Those who use the excrement of chickens reported that it is the best for soil enrichment, because chickens eat all kind of things including worms, snakes, and the faeces of humans and animals. Farmers mostly use it to grow vegetables and fruits.

Reddis collect dung and carry it over long distance to their farm. In the first place, farmers pile up dung, household garbage and ashes together. When dung rot, they will be the best fertilizers. According to informants wet dung is more important than the dry dung because of the fact that the rain makes the dung to drip on the land. As a matter of fact, many farmers smear wet dung on their land so that the land will be able to grow good grasses and crops. For this matter, farmers build cattle pens on different parts of their land. According to the Reddis, using manure is good for yielding crops.

Cattle dung is also useful for fuel. Cattle dung and crop residues replace firewood in the study village. The Reddi women often mix the dung of cattle to start a fire, since it burns well although it does not have the power to stay for a long time. The absence of firewood forces the *Reddis* to reduce the use of dung as a soil enrichment device. Particularly *Reddis* in the village rely on dung as a source of fuel owing to the shortage of firewood in their area.

When the fields become full of holes, farmers use furrow to divert surface runoff to the right and left side of the holes. They put branches of thorny trees and stones in gullies so as to retain alluvial soil that is essential for the regeneration of degraded land. Also, *Reddis* plant various kinds of trees on sloping land to reduce the impact of surface-runoff. I have been informed that *Reddis* in the study area are conscious that once the crops sprout and cover the land, they can retain the sediment carried by runoff and thereby protect the land from erosion. *Reddis* in the study area have identified the following crops, as conservation measures of soil erosion; in order of importance are bean, red gram, and maize etc. *Reddis* also know the water requirements of different crops. Rice, cotton cannot grow well without sufficient rain. Sorghum, *Saggudumpa* (tuber) and tobacco are drought resistant crops, and can grow well with moderate water.

The introduction of chemical fertilizers has had adverse effects on indigenous land conservation methods in the study village. Successive Andhra Pradesh government policies have subsidized and promoted the application of inorganic fertilizers. Farmers lamented that they began to use chemical fertilizers because of the reduction of the number of domestic animals, the scarcity of the land, the fragmentation of the land and the propagation by the government officials about the value of inorganic fertilizers. Farmers are forced to plough their pasture lands and reduce their herds. Animals are either got rid off (by sale or transfer to other persons) or taken to another environmentally richer area. Reddis in the study village, for example, migrate their cattle to other area during summer in order to ameliorate the problem of grazing pastures. Therefore, most of the *Reddis* avoided animal husbandry or reduced herd size. Some farmers also state that the reduction of cattle is partly due to the fact that the bull could not serve cows on time for it could not graze well to achieve the required vigor and strength. The other problem is that farmers cannot take dung from the new villages to their fragmented lands. Thus, they have cultivated all of their plots, steep slopes and marginal lands by avoiding fallowing, crop rotation and dung.

According to Reddis, intensive use of chemical fertilizers has had dramatic negative effects on the land. Unknown weeds have appeared and reduced the yields. Many informants reported that chemical fertilizers multiplied chaffs rather than yields. On top of that, the application of chemical fertilizers has changed the properties of the soil. Most of informants state that the farmland that has been adapted to fertilizers cannot grow any crop without chemical fertilizers. It seems that heavy applications of chemical fertilizers accelerate the decomposition of soil's organic matter and make it unproductive. At the same time, the purchase of chemical fertilizers has become almost impossible for the economically weak small-scale agricultural producers. The price of fertilizers has been increasing over time. From the cost perspective, indigenous conservation measures are cheaper than chemical fertilizers. At present, *Reddi* farmers are in a very difficult situation. They cannot totally ignore chemical fertilizers overnight nor can they rely on dung and other traditional practices because of the stated reasons. Some informants suggest that organic fertilizers should gradually replace chemical fertilizers. The government needs to suspend blanket recommendation of chemical fertilizers. Qualified experts should study the nature of chemical fertilizers. However, some informants state that farmers in some areas have benefited much from chemical fertilizers. Farmers in highlands do not have sufficient cattle to use dung as fertilizer. They have no choice other than using chemical fertilizers.

Division of Labour in Agricultural Activities

Among the *Reddis*, the organization of agriculture activities takes place along gender lines. In rice production activity carried out predominately by men and women, not only in rice production but also in other agriculture activities. Both men and women continue to be actively engaged in agriculture production. Individual knowledge differentiation is visible throughout all aspects of household production. In the study village women are seriously involved in labour work, low agriculture activities and also engaged in activities associated with cooking and domestic purposes. Here in the village women gather minor forest product for food, fodder, medicine and house building as well as for agricultural or other domestic purposes. Women collect minor forest produce according to the seasonal variation in the availability of different products for household consumption. Women have good knowledge of the weeds and pests, which attack crops and the damage they cause. In the village Utlapalem they have specialized roles of men and woman in the management of local agricultural production. Women also make management decisions related to the selection of animals for sale, fodder acquisition and treatment of diseases. In the village women also organize and control rotational grazing strategies to avoid overgrazing. And part of this strategy consists of feeding their animals crop residues or grazing them on fallow plots during dry season. Women also have knowledge of animal health and are able to identify common diseases on the basis of external symptoms. Women bring fodder for domestic animals from the fields and also clean the animal shed every day.

Special Knowledge of Men

In the village, men excel in the following activities: ploughing the field, digging by spade (*Para Pani*), *Akupoyuta* planting of rice crop, *Noorpidi* (Thrashing) taking out rice from the paddy, *Akupikadum* (plucking remain grass), *Kuppasadurkodam* gathering grass of paddy after take out the rice and preparing paddy grass hill, *Pallelu Kotadam,* tubers cutting (*dumpa kotha*), *Adda* or *Gorru tholadam* (ploughing with iron plough), *Tadapu* the small plants of brinjal wet, preparing agriculture implements, weaving variety of baskets the baskets like *Tuttalu, Tadakalu, Gaddelu,* mats etc, preparing *Moku* for climbing palm tree, making the rope, the construction of house roof, spraying pesticides, hunting and cleaning hunted animals, preparing toddy, cutting of palm tree leaves for construction of house and huts, construction of bow and arrow, fencing, harvesting of fruits like tamarind, mango, jack, ambla, etc, selling of fruits and vegetables, bringing pillars from forest for construction, bamboo for basket making, manufacture of bullock cart, preparing of *Gaddi Pandiri,* and *Paaka* in the field, *Kallam chakadam* (preparing thrashing ground), cutting of trees in the forest.

Special Knowledge for Women

According to the male members, they cannot do any thing without women. In agriculture women do *Udupu* (transplantation of rice crop), harvesting of crop is also done by women. Other jobs are sowing seed in the field, helping in many other activities in agriculture. Household work like cooking, fetching water, cleaning of household, smearing, plastering, decorating with *Sudda Muggu*. Bringing fire wood from the forest and broomsticks to make brooms. Women only do *Dumpa* crop plant planted by women and cleaning of animal shed. Women walk in-groups or with their daughters to the forest to collect firewood for household consumption. Sometimes they store the wood on planks of big trees close to the settlement and use it according to their requirements. Fetching, storing and purification of water are done by women only. Some of the crops like pulses are threshed by the women manually with their bare hands.

The work which is dones by combinedly includes weeding *(Kalapu thiyadam)*, harvesting *(Kotha)*, gathering of harvested crop at one place *Panakattadam)*, digging *(Goppu)*, fetching soil though women do it more often for house construction, tuber cutting and harvesting *(Dumpa gita)*. The Appendix 5 provides Konda Reddi division of labour in agricultural activities.

Local Knowledge of Quantification and Measurement

Konda Reddi have their own way of quantifying agricultural produces and measuring lands. Measurement is locally called as *Kolatha*. Length is termed as *Podava*, while *Athu* denotes the height, and *Vedalpu* is the term for breadth. The grains are measured as fallows:

1 *Sola* = ¼ kg, 1 *Thavvedi* = ½ kg, 1 *Rendu Thavvalu* = 1 *Manedi* = 1kg, 1 *Kuncham* = 4kg, 20 *Kunchalu* = 1 *Basta*, 24 *Kunchalu* = 1 *Kata* = 100 kg, 3 *Kata* = 1 *Putti* or *Anu*, 30 *Kata* = 1 *Garse*. They measure lands, sticks, and trees, etc. with the fingers and hands. A measure of four fingers length is called as *Bethedu*. *Nudethedu* is two fingers length, while *Janedu* is five-finger length. Half a hand length, from elbow to the tip of the middle finger is one *Moora*. *Oka Gazam* is the length from the chest to the tip of the middle finger, and *Bara* is two full hand lengths. *Ara Adugu* is the first with an erect thumb, and *Adugu* is the double of *Ara Adagu*. Three (*Mudu*) *Adugulu* is equal to one *Gazam*, and two *Gazalu* is equal to one *Bara*.

Reddis measure the depth of holes dug in the ground, water sources like wells and lakes, etc., with the body lengths called *Kuti Medi*. The measure from feet to knee is called *Mdukulethu*, and the measure from feet to waist is called *Molakattu*. If the water is up to the armpit level then it is called *Sankallothu*, and if it is neck-deep, it is called *Peekalothu*. If the water is up to the full height of a person, then it is called *Neeluvedu*. And Cheyi *Athu* is the depth equivalent to a person with a raised hand.

They measure the quantity of compost according to the quantity, which can be filled in a bullock cart (*Bandi*) and a basket full is termed as *Erodi Butta*. Mangoes and other relatively bigger fruits are counted in *Parakas* and one *Paraka* is equal to 14 mangoes. Other fruits like tamarind and amala are measured in kgs and in terms of sac-full or *Basthas*.

Konda Reddi quantify their fields by acre and cents. 4 Araka is considered equal to 1 acre of land. They quantify hours of ploughing also with 10 hours of ploughing of land being equal to 1 *Gunti*. The measurement of the land and big trees is by *Gazam* (by foot length). Reddis say 100 *Gazams* are equal to one acre of land. For basket making, Konda Reddi take the measurement of bamboos with their hand. A length of half of the hand is called as *Morra*. Different types of baskets are made with different measurements. *Pyada Thatta* (this type of basket is used for carrying animal dung from the animal shed) is prepared with 3 *Moora* lengths of bamboo. *Thatta* (Basket for carrying things to the field) is prepared by 4 *Morra* lengths of bamboo. *Gudu* (basket used to keep the poultry in the night is prepared by 6 *Moora* lengths of bamboo. *Tadaka* (used for fencing the house and also for the construction of bathrooms) is prepared by 10 *Moora* lengths of bamboo. *Gadde* (a big basket used to store paddy) is prepared by 10-15 *Moora* lengths of bamboo. *Ethu* (a small basket) needs 4 *Moora* lengths of bamboo. *Gibbe* (a small basket used as a sieve to filter out the gruel) needs only 1 *Moora* length of bamboo. *Chata* (used to clean the rice) needs 2 *Moora* lengths of bamboo.

Reddis while preparing rope from the petiole of the palm leaves take measurement of the fibre in *Mooras*. *Mooku* prepared for climbing the palm trees needs exactly 5 *Mooras*.

Konda Reddi also have a system of count for the animals. One pair of oxen is called *Kadi*, 2 pairs of oxen are called *Kandlu*, and 3 pair of oxen are called *Kadibandham*. 5 oxen is called *Rendu gandla Bandham*, 11 oxen are called *Idu gandla bandam*. In the case of goats, more than 5 goats are called *dhodi mekalu*. Regarding hen and eggs they quantify in different ways. 4 eggs and 4 hens are called as *Oka punji*.

Social Organization for Cultivation and Crop Protection

Among Reddis household or family form the basic social group for protecting the resources and also for labour pooling. Household as a social grouping continuously work in *podu* fields through out the agricultural season in order to obtain the yield. In addition to the household grouping it is observed that inter and intra lineage pooling of labour power takes place to minimize the time duration as well as hardships associated with agriculture. The members cooperate or assist purely on obligatory lines, but reciprocal attitude in the exchange of labour input is always expected and maintained. Every household or individual feels it is their responsibility to reciprocate in the same form. This obligatory form of exchange of labour cooperation does not give raise to any kind of payments in the form of cash or grains. But the household, which receives the assistance, expresses its gratitude through some form. For e.g. the members of the household serve meal and if possible some Caryota toddy to the villagers who extend their cooperation and work in the field. In such situation the members of the household try to satisfy the members who assist according to their standard of living. Either demanding or bargaining does not take place, since every household at one time or the other utilizes the services of the whole group. Generally such labour assistance is sought at times of *podu* clearings, transplantations, sometimes weeding and harvesting. The Appendix 6 shows Reddis division of labour gender wise.

In the case of crop protection, the members of neighboring fields join together and form a social group. They all collectively guard the crop from cattle, animals, and pests. Sometimes during daytime it is carried out in shift system, but during night time all the members of the group watch the fields.

Rituals or Festivals Associated with Agriculture Activities of Village Community

Generally, *Reddis* plan for the agriculture activities, with the help of knowledge, generated through tradition of experiences, on quality of land, soil, rain and pest etc. but some of them fail in the domain of uncertainty like nature of forthcoming rain and pest attack, because of which they plan for future ventures with caution. In order to overcome this problem, they seek the help of village deities and rituals. The agriculture in the region is completely rain fed, and it is difficult to predict the nature of rain. In order to make a prediction of forthcoming rainfall, the community has devised a ritual of predicting nature of future rain. For the protection of crops, the village deities are propitiated to gain psychological confidence and satisfaction for the communities to cope with the uncertainty of nature. For this, they have designed rituals and ceremonies for various activities. In course of these ceremonies, the villagers conduct some exercises to make metrological prediction like rain and future yield.

The villagers by the way of precaution employ religious rites in conjunction with the activities of agriculture in order that the course of such activities may go smoothly. The agriculture activities begin after the celebration of *Gangalamma* (village deity) festival which they celebrate in the month of May. The villagers state that after this ceremony, the mother earth is ready for production and only then agriculture work will be initiated. The agriculture preparation for the coming seasons begins; one month after harvesting of crops is done of the previous season i.e. after *Gangalamma* festival. The *Reddis* starts ploughing (*Dukki*) their agriculture fields by celebrating *Bhudevi panduga* and leave it to dry off. After one and half month period, the fields are again ploughed. This is termed as *Dukkivarsham* in the local language, which means second or repeat of ploughing. On this occasion straws and other left over twigs and weeds are removed from the fields. The fields are again ploughed third time and it is called *Musurivarsham Dukki* and then it is left over for some time.

For the well being of the crops the *Reddis*, perform rituals at the field. This depends exclusively on various families. The agriculture rituals begin with the performance of *Vithanala panduga* (worshiping of seeds), prior to cultivating a new crop. For this ritual, the farmer takes a hand full of seeds of crop and incense sticks, coconut, and flowers and goes to the field along with the plough and the bullocks. He recites the name of *Gangammatalli* and *Bhudevi* and leave a hand full of seeds in the field and requests the deities that "today I am beginning my agriculture activities by preparing agriculture fields, whose protection is in your hands for which you will be propitiated at the time *Korrakotha* (offering of new grains) festival, therefore it is your responsibility to safeguard the field and its crops". After saying this he releases the seeds in the field and again says to the mother earth "I am putting seeds of crop in your womb, whose protection and growth is in your hands. Hence, please allow me to split your womb to put these seeds into it". After performing this ritual *Reddis* starts their agriculture activities by ploughing the field. The field preparation work is finished by the month of May and June; this is generally done prior to the coming of rain. Next, *Vanadevudu* i.e. God of rain is worshipped for rainfall. Reddis consider that if the God and Goddess were happy with their worship, the yield would be better.

Most of the rituals conducted by the villagers reflect their concern about the production aspect of their crops and have designed speculative mechanisms to foretell the production in future. For this they conduct a ritual called *Korrakotha* for major crops.

Korrakotha: This festival is celebrated after the harvest of new crop like *Samalu, Gantelu* etc. This festival is generally performed in the month of (August-September). This festival is performed to protect the crops from the evil eyes and evil spirits. In this festival sacrifices are offered to village deity. On this day *Goravadu* village *pujari* collects saplings of crops and offers to the deity by chanting mantras.

Varikotha: The agriculture season ends with harvesting and thrashing of the paddy. The produce is collected in the storehouse *(Gadde)* and after offering the new grains to the village deity on *Gangalamma* the paddy is brought into use for eating. This festival is aimed at offering the first produce from the crop to the deities of the village and ancestors who have protected the crop from disease and calamities. Otherwise they will be angry and in future will not protect their crop, but harm the village and its crops. In the rituals, the new grains are offered to the deities requesting them to keep looking after their crops in future in similar manner. This festival celebrated in the month of (November-December) and is celebrated at household level. No specific date is fixed for celebration of the festival. The festival is celebrated with great rejoicing and drinking of toddy and feasting follows this offering of the crop to deity.

Pappu Panduga: The festival celebrated after the harvest of new pulses. This festival is generally performed in the month of (August- September). This festival is aimed at offering the first produce from the crop to the forest God *Bairudu*. This festival is celebrated by village *Pujari*. All the villagers contribute money to buy the chicken and give it to the *Gorvadu/ Pujari* to sacrifice at the altar of God. *Goruvadu* go to the forest to the sacred place along with new crop of pulses, vermilion (*Kunkuma)*, turmeric powder *(Pasapu)*, salt (*uppu)*, oil (*Nunne)*, chilly powder (*kaaram*) and with chicken (*kodi)*. First, the priest cleans the place and give bath to the God with water and applies *Kunkuma, Pasapu*. He then breaks a coconut, lights *Battulu*, touch the feet of God and sacrifice the hen by chanting mantra. After that he cooks the sacrificed hen himself, eats it and returns. This festival is performed to protect the crops from the evil eyes, evil spirits and wild animals from the forest.

To obtain the protection of the gods it is necessary to have a good moral character, refrain from evil acts and make offerings, or *poojas,* and undertake vows. A favour granted by a god has to be honoured with a vow; otherwise the gods may show their wrath in the form of disease, crop failure, drought, or flood.

In agriculture, spiritual practices are designed to protect crops from pests and post-harvest problems. Practices vary from the simple to the elaborate, and are generally combined with other indigenous activities, such as astrology and special consideration for the environment. Traditional spiritual practices are often performed by a professional spiritual healer or shaman. They may include elements, such as repeated sounds, or *mantras*, chanted to get rid of pests and diseases in paddy. While simple practices are freely transferred, the more complicated and powerful practices are kept secret and passed only to certain family members.

Astrology is a major element in Konda Reddi local knowledge systems. Astrology assumes that the arrangement of the planets has a direct influence on earthly beings and processes. Therefore, people prefer certain days of the week for planting root and tuber crops or some types of trees, while other days are considered good for planting cereals, for harvesting, or for working with cattle. During full moon, the farmers prefer to take the first meal from a crop, while other types of field activities are generally not undertaken at this time. Before deciding on a certain agricultural activity, or before a journey, Konda Reddis often consult an astrologer.

The entire agriculture activities like preparation of agriculture fields, manuring of the fields, sowing seeds, ripening of the crops, harvesting and eating of the first produce are closely associated with rituals and festivals. Their knowledge of seed selection and preservation is both extensive and intensive. Collection and preservation of each type of seed needs a separate technique. Preservation is always meticulously done so as to ensure that the seeds are healthy and ready for sowing for the next season.

Konda Reddis who have been influenced by Government programmes, NGO's Programmes and the non-tribals way of cultivation of lands have accepted the use of improved seeds, fertilizers, and introduction of new cash crops. The most outstanding trend observed in the village is the shift from a primarily food based system of cropping to commercial cash crops. Adaptation of modern agricultural practice like using fertilizers and other pesticides etc are very few and still slow to take place in the village. Various inputs such as, tillage of fields, manuring, sowing, weeding, transportation, protection of crops, harvesting, etc., are made in the form of manual labour provided by the farmer.

The output of agriculture is primarily a subsistence output needed to support the farmer and his family in the form of daily food items and his domestic animals in the form of fodder derived from the by produce of the crops raised.

Though cash crops are not grown extensively some cash crops like cotton, tubers (*Dumpa*), tobacco (*Pogaaku*), are grown for ready income. Vegetables are also grown both for consumption and sale. Paddy cultivation is their primary occupation but they cannot produce it for sale as it is enough for their own family consumption, because their land holdings, which range from one to two acres is too small to produce enough food for their family. Other than this, lack of water degrading of paddy fields due to long use and simple farming techniques and methods prevent maximum yield from their cultivated fields.

It, however, does not mean that they do not have profit when their input is compared to their out put. They have profit in terms of their agricultural input and output, but they do not have adequate irrigation facility, and appropriate technology that can support their family economically. The cash crops required more money investments. Debts to banks and moneylenders substantially increased. The pressure to repay loans took a high toll on the emotional and mental well being of the families.

4

Local Knowledge of the Environment and Wild Flora

Evolution of human life and culture has directly or indirectly been associated with nature, and influenced by the surrounding environment. It is well known that the relationship between human beings and the environment in general has never been static. The life, culture and traditions of indigenous communities however, remained static for hundred of years and often it is said that they are the living archaeological museums of ancient traditions and cultural heritage of a nation. Primitive man living closely associated with nature chiefly depended on forests for survival. Human dependence on the plants around him made him acquire the knowledge of economic and medicinal values of many plants and animals by trial and error. Consequently, man became the storehouse of knowledge of many useful as well as harmful plants, accumulated and enriched through generations. This knowledge was passed on from one generation to another, without any written documents. It is therfore important that before this rich unwritten local knowledge on uses of plant resources is lost forever, through the recent accelerated 'modernisation' of the tribals, it must be properly documented and preserved.

Among the several approaches to ethnobotanical research, the most important ones are through archeological study, literature survey, collection of herbarium, and field surveys. The field approach plays a vital role, because of the direct contact that can be established with the indigenous people living in harmony with surrounding vegetation, which can easily bring out the authentic information on the uses of plants both wild and cultivated.

Indian tribes have adapted to natural environments. Their habitats differ as widely as hills, forests, plains, deserts, islands or their combination. Their economy, social institutions, beliefs and practices and in fact their whole subsistence pattern are adaptations to their immediate environment. Therefore, their culture can be better understood in terms of their intricately interrelated nature-man-spirit complex (Vidyarthi, 1973).

Communities living adjacent to forests, who depend on them for their livelihood, have a wealth of knowledge about the use of forest plant products. These communities have lived in and out of the forests using forest resources for their daily needs. They use the forest resources for food production, health care systems, shelter, transport, home crafts and many other needs. They usually have first hand information on location abundance, and management of such resources and methods of obtaining the products. Through the long association with the natural resources, they are the best source of information on what can be utilized. The important forest plant products can be easily identified using local knowledge. This chapter addresses to the knowledge of wild plants and plant products of the Konda Reddi.

Local knowledge among the Konda Reddi is accumulated through observation and real life experiences. *Reddis,* pursue their livelihood directly depending on forest resources. They have sound knowledge of ecology and view the environment as their only source of livelihood, which in different ways influences their lives. Besides, Reddis strictly observe certain social and cultural norms, which regulate and guide the process of utilizing natural resources and aim at conservation practices of natural resources.

This chapter aims at describing the identification, characterization and description of some of the uses of forest plants with the help of local people's knowledge. The chapter focuses on the ethno-ecological knowledge and management of local resources. The study shows how both local practices and native perspectives are capable of forming a basis for sustainable development.

Forest and Konda Reddis

Reddis dwelling in the interior areas have a greater level of dependence on the local natural resources. Undulating terrain, thick forests, canals, streams and hillocks surround the village. Their adaptation to the physical environment is seen in the material culture and in particular technology. Konda Reddis life, whether economic, social, religious and their beliefs and worldviews are influenced by their immediate forest ecosystem. Forest plays a vital role in the survival of this forest dwelling tribe. Forest flora ensures Reddis subsistence, their shelter; inspire their love, their marriage, their recreation and happiness. Trees caution their action and motivate their behaviour. Therefore, Reddis evolved an elaborate system of rituals that bind them with the trees. Their intimacy with the trees is expressed in their folksongs, which they sing for themselves and pass on to the future generations. This deep bond is one of continuity between humans and trees.

The forests not only provide cultivated land for the Konda Reddis, but also virtually everything else that the people need in their lives. Due to the poor conditions of the land, farming among *Reddis* is not very productive and the harvest yields are not good. The people have to depend on forest gathering and hunting to supplement their diet. Supplementary subsistence support is derived from the bounty of nature in the form of seasonal edible fruits, roots, tubers, green leaves, mushrooms, insects, honey, gum, *Caryota urens, Borassus palm* sap etc., throughout the year. Besides, numerous other items of minor forest produce are used to cater to their consumption as well as for commercial needs. The Konda Reddis are able to identify which wild vegetables or mushrooms are edible and which are not. They know which tuber and colorful mushrooms are poisonous. They also know where and when they can find these edible plants. Forests also provide varieties of medicinal plants herbs etc., They also know where in the forests they can find medicinal herbs, that they use to cure their diseases and for other medical purposes.

The human-nature interaction is of primary importance in economic activity of the Konda Reddis and it is intertwined with every aspect of their living. However, this agro-forest based economy is subject to the vagaries of nature. Forest based economy not only provides food but also materials to build huts and other purposeful articles, fuel for cooking as well as lighting and warmth, fodder for cattle, rope, grass lac, soap nut, tamarind, mohwa flowers, beedi leaves, adda leaves, broomsticks, etc. Farming is made difficult due to the natural conditions of the region; the luxuriant, valuable and abundant forests help the natives to survive in the most difficult conditions. They evolved native methods of approach in extracting the forest products for the survival and continuity of the community. In hot dry summers, agricultural activity is meager. People mostly gather forest produce in this period. Forest is the heart of their traditional life and culture.

Different species of wood and non-wood forest products from the forest play a major role in the lives of Konda Reddis. These products are harvested for both subsistence and commercial purposes and represent an important source of livelihood for the Reddis. In the village under study, almost every household is enganged in collection of different wood and non-wood products. Round the year they collect wood and non-wood products from the forest.

Foraging/gathering is an important economic activity. All the members including men, women and children participate in the collection of varieties of forest produce. It depends on the seasonal availability. Throughout the year some or the other forest resource is available. Foraging/gathering is rather a hard and strenuous activity. They start in the morning and go out in the forest in search of tubers and other resources. Due to scarcity, they are now required to travel more into the interiors of forests. Here, again reserved forests are an obstacle for them to collect minor forest produce. The cattle too find it difficult to get fodder or grass.

Their subsistence economy, beliefs both religious and spiritual, health and folk medical practice and social life are all influenced by the ecosystem. Further, for centuries Konda Reddis too with their shared beliefs and practices have been protecting and conserving their immediate environment. They revere and worship forest, streams, rivers, as they are dependent on them for their survival. They strongly believe in spirits and Gods and worship them.

Like other indigenous societies in India, Konda Reddis also depend to a large extent on forest resources to eke out their subsistence. People gather many forest products for food, drink, fodder, medicine and house construction material as well as for agricultural or other domestic purposes. There is a seasonal variation in the availability of different products through out the year for household consumption. Forest products also contribute to the individual household income. In some households, products that are not consumed domestically are sold and the money is used for buying other food items. In others, the money earned from forest products is the major source of income throughout the year. It is also interesting to note that in olden days some of these plants were seen as a kind of 'group identity markers'.

Konda Reddis' Perception about Nature

Reddis carry both kinds of feelings, of reverence and fear towards the forest. They perceive forest as their father (A*daviraju/*), mother (*Adavithalli/Prentalu*) or mountain God (*Konda Devudu/ Birudu*), Goddess (*Kondathalli/Nilaramma*) that means forest is everything for them. Because, forest provides everything for them for their survival and paradoxically it is also symbolized as evil at times, it is foreboding, uncontrollable and dangerous. Reddis think that forest is the habitat of evil spirit and ghosts who attack people when they are in forest. Konda Reddis have mixed emotions of fear, respect and wonder towards the forest and other natural components. The hills and mountains, which are considered to be the symbols of power, are considered to be majestic and mysterious, uncontrollable and threatening. They have the belief that the hills and mountains, which surround the villages, protect the people and villages from natural calamities. Forest plays an important role in the religious life of the people.

Water bodies like streams (*Uta*) and river (*Kaluva*) are revered as the sources of life. Streams are considered to be the village deity of *Gangalamma*. Land too is perceived as mother of Goddess (*Bhudevi*), who ceaselessly blesses them and provides them with agricultural produces. During the time of pollution caused due to the various life cycle occasions such as birth, puberty and death, they cleanse themselves in the stream water, which is believed to make them pure. Menstruating women are prohibited from entering the farms and religious places, as it is considered a sin to pollute it. Reddis perform *Gangalamma* festival to propitiate the goddess of water to have sufficient rains for them to cultivate. In a sense, this festival can be noted as the annual harvest festival where all the harvested grains are ablated to *Gangalamma*. The celebration of this festival is held for three days in the month of March-April (*Shivaratramavasa*), which is celebrated elaborately since the whole village by this time would have completed their agricultural activities. The deity is seated under an old *Burugu* tree located in the grove. A stone is placed under the tree to symbolize the deity. Country chicken and goats are offered every year so as to gain good will. They believe in supernatural powers, deities and benevolent and malevolent spirits. All the religious beliefs, practices and ceremonies are associated with the physical environment.

Forest Categorization

Reddis categorize forest into 4 major types, which is locally known as *Adavi*. They categorise on the basis of location, size, shape and vegetation. The forest on the mountains or small hillocks is known as *Metta adavi*, which characterizes the rocky part of the forest with scanty vegetaion. Thick vegetation on the mountain is called *Pedda metta*. If the forest on the mountain is impenetrable, they name it as *Gobba*.

Konda Reddis also categorize forest (*Adavi*) on the basis of the density of vegetation. A *Kutupu* represents the part of the forest where single species of groves are found. Places with small bushes, which do not exceed knee height, are termed as *Tuppa* or *Donka*. Thorny forest with long bushes is called as *Kayya* or *Tutti*, while parts of the forest with climbers or creepers are known as *Guda*. It is noticed that different *Reddi* villagers use different terms for types of forest, which are highly varied even is small ecological zones. However, what is important here is to note that Konda Reddis categorise forest based on different locally meaningful criteria, which indicates their familiarity with the forest ecosytem.

Moutain or hill is locally called as *Konda*. The big hill is divided into three regions i.e. *Chinnametta, Gubatlametta, and Goulugobba. Goulugobba* is the center part of the hill, *Chinnametta* is the smaller part of the hill and *Gubatlametta* is the bigger part of the hill. The area below the hill and thick in vegetation is called *Chitti Adavi* (small forest). *Banda* is the place, which is open, or the plain area on the mountain. The tip of the hill is called *Koppu*. *Konda addu* is a corner of the hill. *Degubottut* is the slope of the mountain.

Konda Reddis categorize flora into trees (*Chettlu*), plant (*Mokkalu*), herbs (*Chinna mokkalu*), shrubs (Pedda mokka) tendrils or climbers (*Teegalu*), creepers (*Nela teegalu*), tubers (*Dumpalu*), green leaves (*Akukuralu*), mushrooms (*Kokulu*), grasses (*Gaddi*), bushes (*Tuppalu*), weeds (*Donkalu*), bamboos (*Veduru Pidam*), parasites (*Antulu*), Balancing root (Tutlu), Parasite (*Radam* or *Antulu*) etc. Trees are *chettlu*. Trees are further classified into two categories, such as, chettu, meaning a normal tree, and the other one is *Manu*, which means trees having large trunks. Mid-sized plants are called as *Mokkalu;*

Among the Reddis, the size rather than woodiness constitute the underlying distinction between tree and non-tree. They also consider height, width and depth of the tree. The tuber (*dumpa*) and mushroom (*kokkulu*) also assume life-form status for the distinctive role they play in local use by the *Reddis*. At the time of cutting any tree they categorize into the following attributes like *Modalu* (starting point of the tree), *Mayana* (center point of the tree), *Chevara* (ending point of the tree).

The basic division of the living world of the Reddis involves the construction of well-bounded configurations or morphotypes. Reddis describe that a *Chettu* (tree) possesses the following parts like *Veru* (root), *Modu* (stem), *Aku* (leaf), *Letha Aku* (tender leaf), *Muduru Aku* (rough leaf), *Puvu* (flower), *Kommalu* (branches), *Kaya* (raw fruit), *Pikke* (seeds), *Pandu* (ripe fruit) and so on.

Konda Reddis call bamboo as *veduru* or *bongu*. A group of bamboos is locally called as *Pidum* or *katta pidum*. They categorize the bamboo into three types on the basis of their nodes and internodes. The nodes locally are called as *kanupu*.

(1) *Mullem veduru*- This variety of bamboo has thorns at the internodes and the distance between two nodes, is very less. It is most commonly found in the forests. This variety is the tallest among all the bamboos. (2) *Jaripidum*, *jari* means to slip; in this variety of bamboo, the difference between nodes and internodes is large. It is rarely found in the forests. This type of bamboo needs more water to grow. (3) *Sadarnum veduru* or *kondapidum* is a normal kind of bamboo found everywhere in the forests. The bamboo, which is hollow inside, is called as *bongu*. They use this *bongu* for basket making and for mat making. Mat is locally called as *medar*. The hole inside the bamboo is called as *voollum*.

Like the above example, Konda Reddis are able to identify specific plant according to their morphological properties. They identify plants by seeing the structure of leaves, smell of leaves, colour and smell of the bark, sap colour, type of the flower, colour of the flowers, smell of flower, type of the fruits, colour of the fruits, type of seed, size of seed, colour of seed, height of the tree, size of the tree, and the various locations in which the flora are grown. They also classify plants as useful and non-useful, flowering and non-flowering, wild and domesticated, fruit bearing and without fruits, depending on early vegetation and early fruits. Konda Reddis can identify number of plant species, about 90% of them are considered as useful, either for food or for medicine, fodder, or for use in shelter, etc. Their knowledge of flora begins from childhood and is accumulated throughout their life.

The Reddis' classification and nomenclature of plants are based on plant morphology, phenology and use. In some cases, the same name may be given to different plants, although the people do know the differences. Finally, the same plant may be classified and named according to different criteria, resulting in several names for the same plant. Sometimes they have more generic names among herbs than tress and shrubs, reflecting, perhaps, the greater diversity of use of the latter for both human and animal food. Some plants are categorized differently depending on the attributes such as medicinal, toxic etc. A number of plants are named according to their therapeutic value for human and animal diseases. The Reddis of Utalapalem have utilitarian categories for plants, such as shade, perfume, and bedding etc. They have special names for barks of those trees that are used for tannin. The same plant is called by several names corresponding to the early vegetative stages, flowering, and senescence etc. Generally the Reddis names, knowledge and classification of flora is more practical and immediately utilitarian than the formal scientist's. For example, in the village, plants that are neither useful nor good for forage are given the names of useless things or less respected. They consider some plants as beautiful, decorative, unusual or prominent. They have different names for plant parts, such as roots, stem, bark, flowers, leaves, fruits, seeds, branches, tip, etc. Reddis also have an intimate knowledge of the characteristics and values of different plants. For example, they recognize the seasonal availability of different plants and their value in stimulating milk and meat production in livestock, their toxicity to livestock. They also know that certain plants are salty and will deliberately take their livestock to graze on them during the rainy season.

They recognize four types of vegetations depending on their forage value i.e. coarse, tender, salty and poisonous and their effect on livestock constipating, irritating, nutritious, etc. Reddis categorize plants according to the type of soil they grow on and which they are best suited to. All of them will not have an equal amount of knowledge about their local vegetation. For instance older men and women know annual herbs best, herd boys know wild edible fruits best, and honey collectors know most about the flowering of plants. Even within a group, an individual may stand out because of keen powers of observation, prodigious memory, curiosity and intellect.

Konda Reddis Local Knowledge and use of Forest Plant Products

Though agriculture is the main occupation of the Konda Reddis, they are largely dependent on the forest for its various other uses. Reddis gather many food items that are available in the surrounding forests. Foraging is done throughout the year according to the available seasonal forest produces. No concrete boundaries for foraging are maintained. The activity of foraging reveals their intimate attachment with the forest. The following non-wood forest produces are frequently collected by the Reddis: leafy vegetables (*Aku kuralu)*, edible mushrooms (*Kokkulu*), tubers (*Dumpalu*), honey (*Tene*), gums (*Banka*), and seasonal fruits, etc. The list of flora, however, is only representative and not exhaustive. Konda Reddis recognize the following flora, which are available in the forest, with their location and uses.

Reddi's Knowledge of Edible Green Leaves

Reddis identify and collect several edible green leaves from the forest as their food items. *Akukura/Kura* is the local term used for greens vegetable leaves by the Reddis. They identify the green leaves by seeing the leaf stucture, smell, taste, shape and size of the plant. And also identify them by the location of the plants. They are familiar with locations of green leaves plant spread in the forest area and are often knowledgeable about the seasonal availability of a wide variety of these plants. Depending on the season and weather condition, Reddis mostly in groups of four to five persons set out early in the morning or during daytime to collect leaves from the forest in the month of August-December. They look for tender leaves in a plant to obtain good taste and mature leaves are left for the next season. They gather a few leaves from the each plant without uprooting or harming the whole plant. The following leafy vegetables are identified by the *Reddis* in the study village.

Bachalakura plant is a creeper with small leaves and grows near the streams. The leaves are used for curry and soup.

Boddikura plant size is small with small leaves available in the forest. *Reddis* pack these leaves in *adda* leaf, add salt and chilly, roast on the fire and consume. It is a supplement to the main food item.

Bothaboragalakura plants are small in size and are found in the surroundings. *Reddis* prepare curry and consume it as a supplement to the main food item.

Chamakura plant leaves are small in size available in the forest. They prepare curry with leaves and consume it as a supplement to the main food item.

Charykura plants grow abundantly beside the streams in the forest. These leaves are very large and immature. In order to get rid off the putrescense they boil twice or thrice in salt water and prepare curry by mixing with tamarind juice.

Chenchalikura plant is a bushy creeper with small leaves. These creepers are available everywhere near their houses. *Reddis* prepare curry with these leaves.

Dolakondlemkura plant is a creeper with small leaves and grows besides the streams in the forest. The leaves of plant are consumed in the form of curry or a soup.

Goddukura or bhramatantem (Bauhinia sp.) During summer season theses plants are available in the forests abundantly. The size of the leaf is small; they collect tender leaves of the plant for cooking with the pulses as curry and consume with rice.

Gurugukura/Guruvakura (Digeria muricata) plants grow in the ditches, ponds and are also available in wet areas. Both the leaves and petioles are used for preparing curry and soup.

Kokkelikura plants are bushy creepers found near the streams and other wet places in the forest. The tender leaves of the *kokkeli* creeper are relished as a supplement to the main food course by roasting them in adda leaves after adding salt and chillies.

Kondathotakura plants grow in the *chelaka podu* fields in the forest; both leaves and tender stems are prepared as a curry and soup for consumption.

Kondlemkura plants are grown in the forest usually under the shade of large trees. During June-July these leaves are tender and consumable. The plant grows big in size, the size of the leaf is small and sour in taste. The leaves of the plant are consumed in the form of curry, soup and also fried.

Mandamkura plant is a creeper with small leaf and forms a bush in the forest. The plant leaves are prepared as a curry by mixing them with jackfruit.

Pittakandakura plant is a bushy creeper with small leaves available in the forest. Reddis cook these leaves with pulses and prepare curry.

Pullagumadikura plants grow in the fringes of the forest. Both the leaf and stem are used to prepare curry and soup for consumption.

Seekumkura is the tender leaves of sikakai tree available in the forest. Reddis prepare both curry and soup with these tender leaves and the taste of curry is sour.

Thotakura plants are available everywhere near by their houses and fields. The plant leaves are small. Reddis collect them and prepare curry and soup.

Ubasarukura/Ubbakura (*Cardiospermum halicacabum*) plants are found in muddy areas in the forest. The leaves of these plants are small in size. Reddis collect this and prepare curry and soup. Leaf paste is applied on the trunk of the toddy tapping part of Caryota urens for inducing toddy.

Vaskakura is a creeper with small leaves available in the forest. The plant leaves are sour in taste and are mostly consumed along with caryota toddy. The plant leaves are mostly available during the caryota toddy season from December to May. The other leafy vegetables that Konda Reddis consume like *Gonggura, Palakura, kothmira* etc are bought from the market. They identified 22 types of leafy vegetables. Appendix 7 provides different varieties of edible green leaves identified and consumed by the Reddis.

Reddi Knowledge of Edible Mushrooms

Reddis identify various types of mushrooms available in their surrounding forest, about which they possess an intimate knowledge as to which mushroom is edible and which is not. Accordingly, they collect mushrooms from the forest. Their knowledge in identifying different types of mushrooms is extraordinary. *Kokkulu* is the local term used for mushrooms by the *Reddis*. Mushrooms are popular food and they are expert in the art of picking mushrooms. They are familiar with the locations of mushrooms in the forest. Different mushrooms are identified on the basis of their size, colour, taste and smell. Many species of such common edible mushrooms are collected from the forests by the Reddis for consumption. Mushrooms are cut into pieces and are used to prepare curry and soup. The following mushrooms were identified by the Reddis in the study village.

Annekokkulu stem is long and the flower is flat. The mushroom is black on the inner side and the outer layer is white. The stem of the mushroom is attached to the wood and it hangs from it.

Bedi-Bedikokkulu are thick and the colour of mushrooms is white on the outer side and light black inside. The stem of mushroom is thin and is at the corner of the flower, which is flat. These mushrooms are boiled with salt and chilly powder and are consumed.

Budide kokkulu are round in shape with a short stem, which is attached to ground. According to the Reddis all the above-mentioned edible mushrooms are more delicious than animal flesh.

Chirmanu kokkulu flower is flat and big in size, and the stem is thin and small. These mushrooms are slightly red in colour on the cap.

Chiru putta kokkulu is black and white in colour and the shape of the mushroom is round. The stem of the mushroom is slim and the *mogga* is thin. They are hand plucked and packed in the *Adda* leaves. They are roasted in fire and consumed with toddy (*Zeelugu kallu*).

Gadda kokkulu are slightly red and white in colour and are also bigger in size. The stem of the mushroom is stumpy and short and the flowers are small in size. The outer layer of the mushroom is red in colour and the inner is white. *Konda Reddis* roast them and eat them as a side dish along with toddy and liquor.

Gaddi kokkulu sprout on the rotten paddy grass (*varigaddi*) and paddy husk. They are handpicked, packed in adda leaves and roasted on fire and consumed. The mushrooms are black on the inner side and white on the outer side. The stem and flower of the mushrooms are equal in size.

Garuva kokkulu sprouts on the dead plants or trees, so they are named as *garuva*. These mushrooms are cooked immediately after picking them because they are watery and cannot be preserved.

Godugu kokkulu are umbrella shaped; therefore they are called as *godugu kokkulu*. The stem of the mushroom is long and the flower is big and flat. The colour of the mushroom is black inside and white from the outer side. After cutting the mushrooms they are boiled with salt and chilly powder and consumed.

Jonna kokkulu sprout on the decay of *mokkajonna* and *gadajonna*. These mushrooms are bigger than the *putta kokkulu* and the colour is whitish. The stem of the mushroom is stumpy and the flower is flat in structure. They are immediately cooked, as they cannot be preserved, due to their watery contents.

Mamidi kokkulu are big in size. They are hand picked and cut into pieces to prepare a curry or soup. It is mainly used as a side dish to supplement the principle food dish. The size of the stem is long and the flower is flat. The inside colour of the mushroom is black and it is white on the outer side. They consume it in curry and soup form.

Moddukokkulu are big in size. These mushrooms are white in colour, the stem is long and the flower is flat. They are hand picked and cut into pieces to prepare curry or soup.

Nelakokkulu flowers are in round shape, whitish in colour. The life span of these mushrooms is about one month and it takes 2 to 4 days to sprout after the rainfall.

Pandi kokkulu mushrooms sprout on the manure and also in damp places. These are big in size and whitish in colour. The stem of these mushrooms is longer than the flower.

Pedda putta kokkulu sprout up on big anthills, so they are called as *Pedda putta kokkulu. Putta* means an anthill built by termites (*Cheda purugulu*). Mushroom sprouts from depth of the anthill and comes out through the top of it. The size of the mushroom is quite large. They are transparent and white in colour on the top (mushroom cap). They take 2-4 days to sprout. The mushroom has a stem (*Kada*) and the upper part is called mogga, which means flower bud. In this variety, the stem and the upper part are of same size. Immediately after collection, curry or soup is prepared. This type of mushrooms cannot be preserved due to water content.

Pidem kokkulu sprout at the bottom of the bamboo grooves. *Pidem* in the native dialect means a group of bamboos. It starts sprouting with the onset of rains and grows to one and half feet length.

Pidisikokkulu sprout in *borra* or *thorra* (holes of trees). They are medium in size and whitish in colour. They are cooked immediately after picking them, as they cannot be preserved.

Sakaku kokkulu sprout on plain ground. These are larger in size about a moderate palm width and are thick and grey in colour. Konda Reddis cut the mushrooms and boil them with salt and chilly powder to prepare curry or soup. Some times when they collect many mushrooms, they boil them in salt water and smoke them by keeping them on a frame made of bamboo reeds, which is hung over the hearth. This helps to preserve them for a week or ten days.

Veduru kokkulu sprout on the decaying bamboos. They are whitish in colour. It takes 2 to 4 days to sprout after the rains shower. The flower of the mushroom is wide and the stem is small in size. Different varieties of edible mushrooms collected by the *Reddis* are presented in the Appendix 8 followed by the local name and the name of location where they sprout.

Bodda kokkulu are green in colour and small in size. The stem and the flower are equal in size, and the stem is fully covered with the flower.

Burugu kokkulu are red in colour. The stem of the mushroom is attached to the tree trunk and the flower hangs in the air.

Chinta kokkulu are small in size, they are round in shape and the colour of the mushroom is light black.

Dunna kokkulu are red in colour. They are small in size, stem of the mushroom is thin and the flower resembles a flower.

Etha kokkulu is small in size and the stem is attached to root of the tree. The colour of the mushroom is black inside and whitish outside.

Gunde kokkulu mushrooms are whitish in colour. The flower of the mushroom covers the whole stem and the mushrooms look round in shape.

Gupen kokkulu look like a lump and are round in shape. The colour of the mushroom is whitish.

Kovelli kokkulu are red on the outer side and black inside. The stem of the mushroom is long and thin.

Maddi kokkulu are small in size and the colour of the mushrooms is green. The stem of the mushroom is thin and small and the flower of the mushroom resembles a flower.

Mollam kokkulu sprout on the root of *mollam* tree. These mushrooms sprout in a group. The size of the mushroom is small and the stem is thin with whitish colour.

Mudda kokkulu look like a rice morsel, which is called *mudda* in local term. The shape of the mushroom is round and is light black in colour.

Nemali kokkulu are whitish and they are small in size. The stem of mushroom is smaller than the flower.

Oddise kokkulu are red and white in colour and the size of the mushroom is small. The stem of the mushroom is thin and long and the flower is small.

Panasa kokkulu are long and thick and the flower is flat in structure. The colour of the mushroom is white.

Pachoru kokkulu are small and white in colour.

Rai kokkulu are small and are in whitish colour. The stem of mushroom is attached to the tree and the mushrooms are round in shape.

Saama kokkulu are smallest than all the other mushrooms. The colour of the mushroom is violet. The stem of the mushroom is very thin and the flower is also very small.

Tanggudu kokkulu are green in colour. The size of the stem is long and thin and the flower is small.

Tati kokkulu mushrooms are small in size and the stem of the mushroom is attached to the root. The colour of mushrooms is white.

Verri kokkulu are big in size with black colour. The stems of these mushrooms are thick in structure, and the flower is twined around the stem. The word *verri* means waste in local terms.

Vyapa kokkulu are violet in colour. The stem is very short in structure.

The Konda Reddi identified about 19 types of edible mushrooms with equal number of non-eidible varities of mushrooms. Different varieties of non-edible mushrooms identified by the *Reddis* are presented in the Appendix 9 followed by the local name and the name of location where they sprout.

Reddi Knowledge of Edible Tubers

Konda Reddis call tubers as *dumpalu* in the native language. They are familiar with the locations of tuberous plants spread in the forest and are knowledgeable about the availability of a wide variety of these plants. The emic view of forest and its resources are so rich among Reddis that they can easily select a place for digging in search of tuber after observing the herbs and creeper or leaves grown there. They make use of these resources to a large extent. The availability of tubers, wild and semi-wild, which are collected from the forest are used for consumption. The following edible wild tubers are identified by the Reddis in the study village.

Dumpa/kanda/gadda are the terms used for tubers (*Dioscorea spp.*) by the *Reddis*. *Reddis* identify the tubers by the location of plant or creepers, which produce tubers, by seeing leaves of the plants, smell and taste of the leaves and tubers of the plant. They also identify a variety of tubers by seeing the leaves of the plants, size of the tuber plant, size and maturity of the leaves of the plants. Depending on the season and weather conditions, Reddis mostly in groups of four to five persons set out early in the morning or during the daytime. They look for drying plants with ripe yellow leaves to obtain good and mature tubers. They collect the tubers from the soil without disturbing/harming the roots of these plants and also leave a small portion of the tuber back in the pit and cover it with loose soil.

Adavikanda is a creeper that bears pods throughout the year and is available abundantly in the forest. This tuber is very easily obtainable, as it does not need deep digging because they are located very near at the root just beneath the trunk of the creeper. *Reddis* eat it in the form of curry or soup.

Adavitega is a tuber that extends deep into the earth and forms tubers in three or four shelves. *Voluva* and *guanpam* are used to dig under the main stem of the creeper. This tuber is consumed raw and also in the form of curry or soup.

Chauti is a wild tuber. The plant as well as the tuber develops thorns on their outer layers. Each plant has 3 to 4 tubers in different shelves and are dug out with the help of *voluva*. These tubers grow at 2 to 3 meters diametric distant around the bottom of the plant and spread wide under the earth.

Cheda tubers grow on the roots of a creeper underneath the ground. *Voluva* and *gunapam are*s used in order to dig deep and large pits are made directly below the trunk of the creeper. Raw tuber tastes bitter and in order to remove the bitterness, they are kept in flowing water in the stream after cutting into pieces. After this the tubers are boiled properly and curry or soup is prepared.

Donda (Solena heterophylla Lour Cucurbitaceae) tubers are very easily obtainable and are also abundant in the forest. It does not need deep digging because these tubers are located very close to the root just beneath the main stem of the creeper. The creeper bears pods throughout the year, people consume these tubers in the form of curry or soup.

Geetum tubers are dug out using *voluva* and they are cut into pieces and then picked as they are big in size and cannot be extracted intact. Konda Reddis consume both the leaves and the tubers of this plant. These tubers are consumed in the form of curry and soup.

Nara dumpa (Dioscorea wallichi Hook) Dioscoreaceae is a wild tuber available in the surrounding forests of the Konda Reddi habitat. The tubers are located in different shelves that are naturally formed in the earth at different places within the pit. The dead tuber plants are searched and dug with a *voluva* to a depth of about 2 meters around the trunk of the plant. Each *Nara* plant usually has 6 to 8 tubers, each weighing up to 400 to 600 grams. People dig out 2 to 3 tubers, sufficient for the consumption of each family and keep the rest untouched for the next time. In case all the tubers are dug out, one tuber is left behind so that it germinates and provides tubers for the next season. *Voluva* and *chalupu kathi* are used in digging out of these tubers. Since the tubers are not preservable, they are boiled and eaten immediately. When there is scarcity of grains, especially during summer Konda Reddis depend upon these tubers for their survival. *Nara dumpa* is sweet in taste and also the favorite tuber of the Konda Reddis. These tubers are abundantly available and it is processed and consumed as supplement food.

Pindi dumpa or *pidi* tuber plant is a creeper. Its roots extend deep into the earth and form tubers in three or four layers. These tubers are located deep in the earth and need much effort to acquire them. *Voluva* and *gunapam* are used to dig these tubers out. These tubers are consumed raw and also in the form of curry or soup.

Putta Kanda (*Amorphallus paeonilifolius*) *Dennst* (*Nicoloson* (*Araceae*) tubers are boiled with leaves of tamarind, sliced and washed in running water. The washed slices are boiled with fruit juice of tamarind and eaten by Reddis.

Racha vymu tuber is raised by the Konda Reddis in their kitchen gardens located either in the back yard or in front of the dwelling. These tubers can be obtained only after three years of the plant growth. Every year the plants germinate during June to July, grow and appear to die during summer. This happens for the second and third consecutive years.

After the third time of its death people dig it out and consume it raw or prepare curry. These are the largest tubers that are available in the forest and each tuber weighs around three kgs.

Sare tubers are available nearby water pits and streams. The tubers are obtained during January to May. *Voluva* is used to dig out the tuber. *Konda Reddis* consume both the leaves of the plant as well as tubers. In order to get rid off the putrescence that is found in the tubers, they are boiled properly and consumed, in the form of curry and soup.

Savida (Dioscorea puber Blume) tubers are available through out the year and it needs a lot effort to acquire them. *Voluva* and *gunapem* is used in order to dig deep. Large pits are dug directly below the trunk of the creeper. The raw tuber taste bitter and in order to ridit of its bitterness it is kept in the flowing water in the stream after cutting into pieces. For 3 to 4 days it is boiled with salt at regular intervals of time. Curry and soup are prepared after adding tamarind juice, turmeric and chilies etc. The tender tubers tastes more bitter than the developed one, so only mature and fully developed tubers alone are collected. It is an indication that the tubers are fully developed when the large leaves of the creeper turn yellowish. At least two persons will have to labour to obtain this tuber and mostly Konda Reddis prefer to forage for this tuber during their leisure.

Teega /Thega (Naravelia zeylanica L.) DC. (Ranunuculaceae) tubers grow at 2 to 3 meters diametric distance around the bottom of the plant and are spread widely in the earth. This is a wild tuber and both plant and the tubers develop thorns on their outer layer. Each plant has 3 to 4 tubers in different shelves and it is dug out with the help of *voluva*. They consume it as curry with rice.

Vayamu (Dioscorea pentaphylla L.) is a wild tuber and the plant as well as tubers has thorns. These tubers are quite large and weigh about 2 to 3 kgs each. These are dug out after the death of the plants. Konda Reddis believe that unless the plant dies, the tuber is not completely developed. While the tuber is collected a small piece of tuber is left behind for further germination. The tuber is cleaned properly by keeping it for 2 to 3 days in the water to get rid off the putrescence as well as the bitterness in it. Later, it is boiled for a long time at two different intervals of time and is consumed by adding salt and pepper.

Velichanda Gadda ((Dioscorea hispida Dennst. (Dioscoreaceae) sliced tubers are kept in running water for a day, then boiled with leaves of tamarind water is decanted, then cooked and eaten by the Reddis. Appendix 10 provides different varieties of edible tubers identified and used by Reddis.

According to *Reddis*, fully matured, big, ripe and tastier tubers are available between April-May, October-December and January-March, and the rest of the year the tubers are not tasty or fully mature. Before consumption of tubers, a long process is involved in removing the high content of toxic components. Sliced tubers of *Nalla Gadda, Nara dumpa, Vayanu dumpa, Velichanda Gadda* are kept in running water for a day. After a through wash, they are boiled with leaves of tamarind; the water is decanted, then cooked and consumed by the Reddis. Among tuberous plants, some tubers are also used as medicine by the Reddis.

Tuberous plants possess a rich diversity in this region. Genetically diverse crop varieties enable Konda Reddis to enhance the food security of their household and to exploit a range of crop products. In additon, some tubers have therapeutic value and are used in tradional medicine by Reddis. This is an important resource with tremendous future prospects.

Reddis identified 16 varieties of edible tubers. Depending on the season, Reddis engage in food gathering activity. The forest provides numerous edible fruits throughout the year to cater to their consumption. They eat several edible fruits in summer from the forest while they are working in the forest and fields. Fruits collected by the Reddis are presented in Appendix 11, followed by the local name and the name in English.

Local Knowledge of Mango Trees

Reddis identify different local varities of mango (*Mamidi*) trees found in the forest. Only a few elderly people can recall these types of mango trees that were identified and named by their ancestors. According to them, only seven types of mango trees are now found and the rest are extinct. The seven types of mango trees found in the forests and in their habitats now are *Vennalamamidi, Mekasanlamamidi, Kothiperralamamidi, Kadatalmamidi, Boddu mamidi, Metta mamidi,* and *Madammamidi.* Among all these varieties, only *madam* and *boddu mamidi* are sweet mangoes. Wild mango trees commonly found in the forests are called as *Kondamamidi.* Reddis categorise mangoes and mango trees according to their location, shape and size of the fruits, period taken for ripening of the fruits, taste of the fruit, quantity of juice in the mangoes, and the quality of the juice. *Konda Reddis* identify at least 25 varieties of mango trees in the forest. The mangoes and their local names are: *Yennala* or *vennala mamidi, Gedala mamidi, Chidi mamidi, Munta mamidi, Meka sallu mamidi, Varad mamidi, Kothiperrala mamidi, Thokkalamamidi, Kadatal mamidi, Meka burrakayala mamidi, Yelugudu mamidi, Madam Mamidi, Boddu mamidi, Peesu mamidi, Goppu mamidi, Purugulu mamidi, Karre mamidi, Rati mamidi, Metta mamidi, Pendli mamidi, Bangaru budde, Goa mamidi, Kobbera mamidi, Kukumu mamidi, Kodavala peedi mamidi* etc.

Wild Plants used for Extracting Local Drink

Toddy is extracted from mature trees of thati chettu, jeelugu chettu, and etha chettu throughout the year. It takes nearly 20 years for a plant to reach maturity and become ready for extraction of sap. In summer season, Reddis collect Ippa flower (*Madhuca latifolia*) which are either used for making drinks or sold in the local markets. Plants used for tapping toddy, extracting liquor and used for other purpose are presented in the Table 1 followed by the local name and the name in English.

Palms are important trees among Reddis. They form a vital component of forest and agricultural ecosystems, providing a wide range of economic products necessary for daily life. Beside commonly cultivated palms like coconut, wild palms

Table 1
Plants used for Tap Toddy and Prepare Liquor by the Products

Local name	*Common and Scientific name*	*Tree part used for*
Thati chettu	Palmyrah urens/ *Boraruss palm*	Sap used as toddy
Jeelugu/jeeguru chettu	Caryota/ *Aeschynomene indica*	Sap used as toddy
Etha chettu	Phoenix sylvestris	Sap used as toddy
Ippa chettu	Madhuca longifolia var latifolia	Flower used for making liquor
Vepa chettu	Azadirachta indica	Sap is used for toddy
Bodda chettu		Sap is used for toddy
Tumma chettu	*Acacia Nilotic*	Bark is used in toddy and liquor
Somi chetttu	*Soymida febrifuga*	Bark is used in toddy

such as Boraruss *palm* (Thati chettu), Caryota/ *Aeschynomene indica* (Jeelugu chettu), Phoenix sylvestris (Etha chettu) are abundant in the area.

Fish tail palm or *caryota urens* is a palm common in the plain and hilly forest areas of Andhra Pradesh. This palm usually grows near villages and is popular for its sugary sap. The palms attain 25m heights, the trunk has scars of fallen leaves; leaves about 6m length. Konda Reddis commonly call this as *Jeelugu*. After attaining fullest dimensions (usual height of 40 to 60 feet and a base girth of about 3 to 4 feet) the trees give-out first flowering spathe, which appearing at the summit, and stays through the tree's further growth in height. A healthly vigorous tree generally has 15 to 20 fronds of leaves, and once flowering begins, a spathe is let out at the axil of each frond. As the frond travel downward, the spathe increases in size until the middle of the crown, where the largest is given off. Thereafter, the fronds given off are smaller in size and emerge at longer intervals and in a year or two after the last and smallest cluster mature, the tree itself speedly declines and dies. The caryota palms are short lived and the usual life span of the palm is about 15-20 years.

Among the Reddis, the caryota palms are used for tapping the toddy and this is by far the most important product. Toddy is delicious, milky and like water. The sugary sap itself, cool and refreshing. It flows-out at the cut end of the flowering spathe. The *Phoenix sylvestris* (Etha chettu), *palmyra urens* (thati chettu), and the *caryota urens* (Jeelu chettu) undergo distinctive method of tapping. Perhaps, the most laborious yet interesting system that is follwed in the extraction of the toddy from the gigantic flowering-branches is situated at the end. The first spathe let out is usally smaller in size and ill formed and is not tapped. In a month or two, the second spathe appears, and the tapping is started. A couple of bamboos with portions of the arms retained (side branches on each side to serve for steps) are tied on to the stem of the tree, one above the other, untill they form a ladder long enough to reach the spathe. At the top, directly below the spathe a rude platform of a little sticks is constructed for the tapper's use. The tapper ascends the tree and perched on the platform, proceeds to cut away that portion of the tough sheath, which encloses the long, tender flower-spikes of the spathe. The remaining part of it protecting the peduncle of the spathe itself is allowed to remain as a shield from direct heat of the sun or any other external influence that affects it. The spathe, at the time the sheath is removed would have just completed its elongation, but not yet burst open. On removing the sheath, the delicate spikes of unexpanded flowers are all unraveled and stretched. These are wrapped up in few stripes of banana leaves and tied around with a little coir rope.

Early, the next morning, the spathe is untied and the spikes of embroyonic flowers are taken, one at a time, held between forefinger and thumb and each in sucession is gently, yet firmly pressed. This is the first step in training the spikes and it continues for about 5 days. On the sixth day after training, the sheath is cut further to 8 inches approximately. The dense rusty tomentum is scrapped together and ground, with a sour toddy and made into a stiff paste. The paste is applied on the peduncle and this is said to be useful in arresting development of fibrous tissues and decaying in the spikes. For the next 2 or 3 days the spikes, placed between the palms of both the hands and are gently rolled forward and backwards for several times. By the time operation ceases, the fresh toddy begins to flow. This toddy is not collected as is it is astringent and slightly sub acidic. On the third day, the pot, usually an earthen vessel is tied on to the cut end of the spathe and peaces of leaf stalks are tied all round the rim. This is mainly to prevent the rats, squirrels, bats, flying foxes from preying on the contents. Monkeys are however reported to forage on the contents and damage the earthen pot. Every morning and evening the tapper climbs the palm and empties the toddy into plastic containers. Thus tapping is continued for 2-3 months from spathe. Inorder to increase the fermentation and taste of the toddy, roots of *Bauhinia racemosa* (Ari), bark of Cassia fistula (*Rela chekka*), and *Strychnos nux-vomica* (*Musti chekka*) are added. When the tapping is going on a particular spathe, the others do not so readily emerge. Some vigorous palms yield copious amounts of toddy and the pot is to be watched and emptied several times in a day. An average sized spathe is tapped in about four months, and it usually takes about 2 years for all the spathes of the palm to be tapped. The freshly collected toddy is sweet and fairly large quantities can be drunk without consequences. When once fermentation sets in, it soon turns into a foaming, beer-like liquid, acidic in taste and intoxicating readily. The fresh toddy is wholesome nourishment to the Reddis in a family from a 5-year-old child to an old aged grand person. Sitting and enjoying the toddy is a common view during the season.

Phoenix loureirii is an important component of forest flora commonly known as hill date (*Konda eetha, Chitti eetha*). The palms are stunted and gregarious and about 1.5 m in tall, stem is stout and short, covered with persistant leaf sheaths. The palms resemble the *Phoenix sylvestris* (wild date palm) in morphological features except short staturedness. Unlike wild date, the palms are not suited for toddy tapping. Flowering and fruiting of palms is generally observed from March to May. The drupes are 1.5 cm long, oblong, ornage to black when ripe and seeds grow vertically (Murali, *et al*).

The great ashy green fronds of leaves are often used in decorating pandals and stiff dark fibres are used as fishing tackle. Its huge trunk which, when dead and hollowed out, is made into gutters and troughs. Its soft, inner pith is used for sago manufacturing.

Multipurpose use of Borassus Palm (tati cheetu): Konda Reddis named this tree as Tatichettu and they use this tree and its parts for multiple purpose. This plant takes nearly 15-20 years to reach maturity. After sowing the seed *(Tenka)* it takes one year for the seed to sprout and it sprouts with one leaf. When the seed starts germinating, it splits and cotyledons first come out, which are locally called as *Tenka pappu* in the fourth month. In the seventh month the main root grows, which they call *Tati tega* and Reddis collect these roots to eat. But at the stage of *Tati tega* the dying plant is collected for eating. After three years people start using the fibre from the petiole of the leaves as rope things. After ten years they start taking out *Balage* for *Peesu* and also use leaves for construction of houses roof. But they do not use fibre for the construction of house because it is not very strong. After fourteen years, inflorence (*Moovu*) comes out which is a sign of maturity. Konda Reddis eat this *Moovu* raw, but if had in excess, it leads to blood motions and *Vaatam* (paralysis). If the *Moove* is cut out or removed, the tree dies. People extract toddy from this *Moovu* only.

After twenty years the tree starts giving fruits. Toddy can be extracted from it at this stage only. When fruits ripen Reddis also extract toddy from the fruits. But drinking this toddy generates heat in the body. Leaves of this plant are called as *Matta,* which are used for the purpose of thatching of houses, animal shed (Godla paka), field shed (*Chenupaka*) in the fields, fuel for cooking and for other purpose. Reddis use fibre from the petiole of the leaves as a rope in house construction, fencing house and fields. Reddis prepare *moku* from this fibre, which they use for climbing palm tree for toddy extraction and cutting leaves. This *Moku* is also sold at Rs. 50/- per each. Fibre locally called as *Tadapa* (upper layer) *or Nara* (inner layer). They use Tati Tadapa for tying plough and bullock carts at the time of manufacturing. This Tadapa is also sold at Rs. 15/- per hundered. Tree fruits are edible, tender fruits are highly relished which they locally call *Tati Munjalu*. Tender sprouts of these trees are roasted and consumed.

The petiole of the leaf, which has a broad structure, is locally called as *Balige*, from which they take out fibre, called *Peesu*, which has commercial value in the market, and this *Peesu* is sold at Rs. 8-9/- per Kg. In the study village Konda Reddis sell *Tati* leaves at Rs. 40/- per hundred. Extracted toddy from the mature trees, which everybody consumes in the evening, is sold at Rs. 5/- per chambu (half litre).

By tearing the *Tati* leaf, *Reddis* take out *Reka*, which is used to make temporary baskets for carrying harvested crops, vegetables, fruits, and other things. They also use this to weave mats from small size to large sizes. For immediate use when they need to carry animal flesh or some thing in small quantity, they instantly prepare *Mongam* with a leaf. They prepare *Rekha* with the leaf to drink toddy and water in the fields or in the forest.

People use *Tati* wood for the construction of house roof and Pillars. The inner part of the wood is called *Modu* and the upper part is called *Moddu*. The inner part of the trunk is called *Bondu* and the upper part is called *Moddu*. The root of the tree is about one meter long under ground and is very hard. They also sell the wood to other people at Rs. 500/- each. *Bondu* of the tree, the innermost part of the tree trunk is dried and pounded into powder and filtered, with which they prepare *Ambali* (gruel) for consumption. With *Tenka Gunji* they prepare curry and the water of the *Gunji is* used to heal small boils. The petiole of the tender leaf is burned in the fire, mixed with water and the water is used to cure *Chevi potu* (ear pain). Reddis say that earlier they used to prepare slippers with *Tati balige* and also umbrellas with the leaves. Now also in rainy season some people use the leaves to protect themselves from the rain. Reddis also prepare *GaliPatam* (fan), flute, mats, baskets, caps etc. with the *Tati* leaves.

Fish Poisons

The fruits of trees presented in Table 2, are made into a paste to kill and catch fish.

Table 2
Plants Parts are Used for Fish Poison

Local name	*Scientific name*	*Part used for*
Odise	Cleistanthus collinus	Fruits used as fish poison
Indhumu	Strychnos potatorum	Fruits are used as fish poison

Cosmetic Plants

Forests also provide varieties of cosmetic plants to cater to their hygiene. Reddis identify and collect cosmetics for their use as well as for commercial needs. They use Sikum (Acacta sinuate), Kunkudu (*Sapindus emarginatus*) and *Yelakkaya* for head bath. *Yelakkaya* tree fruit is used for dandruff alleviation. After collecting fruits they allow it to dry for a few days. Dry fruits are crushed properly and then mixed in water for sometime and then applied on the head while taking bath. In the village, many people use this as a shampoo and it is said that it is better than the shampoos available in the market. According to Reddis, taking bath with these herbal products makes their hair strong and black. Plants used for cosmetics are presented in the Table 3, followed by the local name and the name in English.

Table 3
Plants Parts Used for Cosmetics

Local name	*Scientific name*	*Plant part used*
Sikum	Acacta sinuate	Fruit in summer season
Kunkudu	Sapindus emarginatus	Fruit in winter season
Velagakkaya/ Yelagakakaya	Limonia clephantum	Fruit in rainy season

Identification and Classification of Grasses

Various types of livestock play a vital role in the agriculture-based economy of Konda Reddis. Goats, chicken, cows, oxen, buffaloes and dogs are found among them. A great variety of grasses and plant products are annually harvested from the forest (presented in the Appendix 12 followed by the local name and the name in English). Fodder is another plant product collected from the forest by Reddis for their livestock. Depending on the season, different types of fodder are available. The knowledge of fodder in the forest is so rich among these people that they can easily select which herbs, creepers and plant leaves are better as fodder. Leaves and twigs of plants mentioned in the table are especially collected (cut and carried) for their domestic animals.

Reddis categorise grass based on useful and non-useful, wild and domesticated, edible and non-edible. The identification of the grass is based on size and colour of the grass, location where the grass grows. The following types of grass are identified by the Reddis in their region. Appendix 12 provides information regarding variety of grasses and plant products used as fodder.

Yerra shangala grass grows in dry lands. The stem of the grass is red in colour and it has 4-5 leaves. The grass does not have tuber like structure at the root part.

Pala shangala grass is also located at the same place where *Yerra shangala* is located. The difference between these two grasses is that *Pala shangala* grass has white colour stem (which they locally called *Pala Rangu* (white colour).

Boddu thonga or Pedda thonga grass has tuberous root, which is nearly about half Kilogram. It gives different types of flowers, which resemble a naval (*Boddu*) in shape and the flowers are black, red and green in colour. With the onset of the rains, this grass starts growing and after three months it flowers.

Jeebu thunga grass grows dense especially in the fields. Red colour flowers and the roots of the grass are in a cluster of about hundred small tubers, where each tuber yields a leaf.

Nara tunga grass grows with single branch and it has tuberous roots. Grass stem and flowers are thin and green in colour. It mostly grows in the rainy season in the agricultural fields and is available where there is water. Because of the tuberous root it is named *Nara tunga.*

Garki gaddi grass has white flowers. The grass has nodes and inter nodes *(Kannalu)* and the roots have many root hairs. *Reddis* use this grass as fodder for cattle and according to them the grass is consumed by donkeys and horses.

Kanpa gaddi grass is very strong and looks like sugarcane. In the study village, this grass is found at the corner of the canal. It has nodes and inter-nodes on the stem. The nodal region is called as *Kanapa* and therfore it is named as *Kanapa gaddi*. The grass has green flowers in winter season. This is the tallest among the grasses. The stem of the grass is very thick and it grows very fast.

Neeti arika grass resembles *Garki Gaddi* and is found in the water and therefore named as *Neeti arika*.

Yenugu Mokalu grass is found in water and floats on the water surface. It has nodes and inter-nodes at a distance of three fingers and each node has leaves.

Aggiantapu kada grass grows along with paddy crops if water is available and it has leaves to its branches.

Madunu Kada grass resembles a groundnut plant with single root and root hairs.

Eelaka chevi Kadag grass leaves resemble the ear of rats and therefore named as such. It is a creeper and has many nodes with leaves.

Jerri gaddi grass is a creeper and spreads on the ground. Once taken out the grass cannot grow again. Domestic animals like to eat it as it is very tasty. The grass is propogated by a seed type *puthika*, which sticks to the body of cattle.

Pyru gaddi grass is a creeper and spreads on the ground. A special character of the grass is that the number of leaves at the nodes of the stem is the same as the number of root hairs at the nodes. It is edible by domestic animals and the grass grows in green grass fields like *Gantellu, Sollu, Jonna, Nuvvu*, etc.

Chepala Gaddi grass is similar in all aspects to the above grass *Pyru Gaddi*. The only difference between them is that the leaves of the Chepala *Gaddi* are short in size.

Gokkeri Gaddi grass leaves are thin and long with red spots at each node and the special character of the grass is that it grows very fast because its lifespan is very short,i.e., three months.

Nulaka Gaddi grass is available for only four months from rainy to early winter season. It is wild grass, found in the forest. Cattle eat this grass when it is tender. Fully matured grass is used to make ropes and bullock carts.

Dabba Gaddi grass is found in *Boddu* (in between the rice fields, they make *Boddu* to supply water from one field to another).

Rella Gaddi grass is green in colour with white flowers. The leaves are very sharp and can cut the hand.

Chepuri Gaddi grass when fully developed becomes hard and is used by the women to make broomsticks to sweep the house. Domestic and wild animals eat *Lady Gaddi* grass in the winter season. Reddis identified 22 varieties of grasses and highlighted their characteristics.

Collection of Honey from the Forest

Honey is called *Tene* in the local dialect. The collection of honey is another important foraging activity of the Konda Reddis. Honey is available for about six months of the year except during the rainy and early winter seasons. Generally, March, April and May is the most productive period for honey collection. Mostly people identify and observe beehives from the very beginning of the formation. Before the beginging of monsoon, Reddis engage in honey collection though they are engaged in collection of wood and other products. They always follow the local calendar to get the best result of harvesting honey. From their experience, they know that honeybees preserve nectar in their colony for a certain period of time and bees consume the nectar themselves in the new moon. Therefore, harvesting must be completed before the new moon. Entering into the forest, they can exactly identify the productive colonies. It may be pointed out that, Reddis do not work on individual basis unless it is small hive. To collect honey from very big hives they form group of between 5-10 men who work cooperatively in their honey collecting activities. The honey collector uses gunny bags to protect oneself from the bees. They roam far away for their village and spend much time for searching the beehives. Generally they identify the location of beehives by seeing the bees around or by smelling the honey.

First of all, Reddis try to dislocate the bees from the colony by *rotte* (a stick with a bundle of dry and wet leaves on the head of the stick, which produces smoke towards the beehive to drive away the bees. Then, they carefully separate the part of colony containing honey and keep it in a pot. This action is always done with a higher sincerity to avoid the destruction of bees and colony forever. They believe that a colony can be used for further hunting if the queen bee and her surrounding associate members remain with her in the colony and are protected. After the collection is over, they either take it home or crush it there itself to extract the honey, and finally preserve it in a glass bottle, plastic jar or pots. They say that only the honey made from the flowers of the domestic plants like food crops tastes good, and if the nectar is collected from the flowers of forest trees like *yerrisenda, kurrasenda, ganuga puvu* the honey does not taste good, rather it tastes sour (*pullaga*) and bitter (*chedu*). The structure of honeycomb is called Bugga and after the honey is removed it is called as *atta* and the honey is called *Nunee*.

Reddis categorise the beehive into three parts. The upper part of it is called as *Tene* (the honey is stored in the upper part of the wax comb), middle part is *pupam* and lower part is called as *pillalu*. The *pupam* is built with secretions from the legs of the bees. According to their knowledge honeybees take 10-20 days to fill the comb with honey. Konda Reddis consume the honey as well as preserve it by heating. Though they do not use honey very regularly they store it for periodic consumption.

Konda Reddis mainly distinguish three types of honey i.e. *chettu tene, torra tene, putta tene*, which are categorized and named on the basis of location of the honey.

1. *Chettu tene*: where bees form their hive on the branches of the trees.
2. *Torra tene*: where beehives are formed in the cavities in the tree trunks.
3. *Putta tene*: where the honeycombs are formed inside the anthills.

The other categorization of honey according to the *Konda Reddis* is based on types of bees, size of the honeycomb, size of the bees, shape of the honeycomb, behaviour of the honeybees, colour of the honey, colour of bees, taste of the honey, taste of the bees, etc. The types of honey that Reddis collect are (1) *Terra thene* or *pedda thene* (2) *Borra tene* (4) *Mosari tene,* in *mosari tene* according to the size of the honeycomb and bees, *Reddis* categorized three types of honeys i.e. *pedda mosari* (big), *bolli mosari* (medium), and *china mosari* (small), (4) *Kundakanduru tene,* (6) *Pacha kanduru and* (7) *Kukkanaluka tene.*

The *chettu tene* and *torra tene* are available from the month of November till the end of summer. *Putta tene* is abundantly available during early rainy season. Konda Reddis while collecting all the above types of honey, use tools such as digging stick, long bamboo, *chelapu kathi,* gunny bag (to cover the body from the attacking bees), basket, tin, vessel, pot to carry the honey.

1. *Terra tene* or *pedda tene*: The beehives are big in size. The size of the honeycomb is is big and long in structure. The colour of the honey is light reddish and it is very sweet in taste. The sting of these bees cause fever. These beehives are seen often on the branches of big trees like, *chirumanu* (*Anogeissus latifolia*), *pacheri, burugu, ellora* etc. The quantity of the honey in the beehive is more and it is not very viscous. *Terra tene* is mostly available during the rainy season. In summer season the honeybees hatch their eggs in the compartments of the beehive. An exceptional colony of big honeybees can produce up to 20-30 kgs of honey. *Konda Reddies* say if *terra tene* is eaten more it generates heat in the body.
2. *Borra tene*: These honeycombs are formed on branches of the trees, cavities, rocks etc. The honey is brownish, viscous and sweet in taste. This variety of honey is mostly available in rainy season. The honeybees are in medium size and their bite is considered to be dangerous. The cavities of the honeycombs are called as *kanna* and they are red in color.
3. *Karra tene*: This variety of honeycomb may be formed on the branches of any tree. They are also found in *tupalu* (bushes), *pidem* (group of branches). Both the beehives and honeybees are small in size. The quantity

of honey is less, light red in color and is sweet in taste. The bite of these bees also causes irritation and pain.

4. *Mosari tene*: These beehives are found on the branches of trees, *pidem* (bamboo groves), rocks (Pedda rai) and anthills (*putta*). This honey tastes sweet. The bees are red in colour. This variety of honey is further classified based on the size of the bees. Big sized honeybees make *Pedda mosare tene*. Medium sized honeybees make *Bolli mosare tene* and small sized honeybees make *Chinna mosare tene*. *China mosare* honey tastes sweeter than the other two. These small sized bees are not aggressive and they rarely attack and bite. The quantity of honey varies according to the size of the bees and their hives. In *pedda mosare* the quantity is more, *Bolli mosare* honey is less and in *chinna mosare* it is little. All the three varieties of honey are available in rainy and winter seasons.
5. *Kundakanduru tene*: These are found in the anthills. The honeybees are big in size and are black in colour. The shape of the honeycomb is like a pot and therefore it is called as *kundakanduru tene*. *Kunda* means pot. The bite of these bees is dangerous and causes fever. The honey is viscous and sweet and it is mostly available in the rainy season.
6. *Pacha kanduru*: These are found more in the bushes (*tupalu*) *gote*, *thati* leaves and *Sandra* trees. The bees are green in colour and therefore known as *pachakanduru*. The size of the honey and honeybees are small. The quantity of honey is less and it is also not very viscous. If the bees bite, the bitten place becomes red and it pains. The taste of honey is sweet and it is mostly available in the rainy season.
7. *Kukkanaluka tene*: These honeycombs are found on bamboo bushes (*pidem*), *kampu donka* (bushes), *gote manu, chirikemanu, and usari manu*. The shape of the beehive is like dogs tongue that is why it is called as *kukkanaluka tene*. The bees are small in size and their

sting causes red rash and irritation. Honey is sweet in taste; but the quantity of honey is less and it is not viscous. This variety of honey is available in the rainy season.

Collection of Gums

Konda Reddis locally call gum as *Banka* and categorize gum into three types as:

1. *Thellati Banka*–White gum.
2. *Tarudu Banka*– Brown gum, and
3. *Nallapu Banka*-Black gum.

The trees like *Kovellichettu, Konda Goa* or (*konda Gogu), cherimana, Maddi,* and *Gumpen* secrete more gums than other trees. To extract gum *Konda Reddis* use *geetha kathi.* Gum is available throughout the year, but the gum secreted in the summer season is white in colour, more in quantity and is of good quality. In the rainy season the gum turns into black colour, because of the absorption of water. For extracting gum, the *Konda Reddis* first make an incision in the bark of tree with *Geetha kathi* on one side. The place where they cut is called *komma.* After this, within three days the secretion of the gum starts. They do not make incision in the same place next time as the quality of the gum may decrease.

Among these three types of gums, the white gum fetches the *Konda Reddis* better price than the others, as it is considered to be of superior quality. GCC pays them at the rate of Rs. 120/- per kg of white gum, Rs 80/-kg for brown gum and Rs. 60/- kg. for the black gum.

Their activities related to non-timber forest products (NTFPs) have been confined to collection, processing and marketing. Some endangered NTFPs are listed in the Appendix 13.

Forest Plants for Construction Purpose

For centuries Reddis have followed a traditional method of construction, using locally available material. This is part of their traditional knowledge base. Building methods and organizational structures are determined by traditional beliefs, norms and values and availability of materials.

For the construction of house, manufacturing furniture, making agricultural and hunting implements *Reddis* make use of the trees found in the forest. Different trees are used for different parts of house construction. Construction wood is used in the form of poles, beams and roof for houses, huts, fences and shelter. They use particular trees as central pillar (*Rattalu*), attics or side poles (*Pattellu/sajja*), beams (*vasallu* for vertical support *or dullam* horizontal support) etc., and they use different types of fibre for tying, bamboo is used for making walls, roof, fencing etc. They utilize the locally available forest products for the construction of house. In their vicinity, *Borassus palm* (Thatichettu), grow abundantly. They use *thatiaku* (*Borassus palm* leaves) and *Dakkagaddi* for thatching roof. They are arranged compactly and are put in many layers. They start tying Thati (*Borassus palm*) leaves from one corner of the roof. The first row starts with the lower side of the leaf and at the top of the roof the leaves are covered on the upper side of the leaf, which lets the rainwater run off. This technique of covering leaves of *thati* tree prevents the seepage of rainwater. *Konda Reddis* dismantle the old roof and make new roofs every three years.

Traditionally, men are responsible for constructing the wall and roof frame, thatching the roof and making the sun dried mud. Women assist the men by collecting leaves for thatching, plastering the walls, and preparing the mud and dung surface of the floor. They decorate their houses with a mixture of *yerramatti* (red soil), sand and dung on the walls. In the front and backyards, they smear (*aluku*) dung water and then decorate with lime powder (*muggu*). The Appendix 14 provides information about the Plants used by the *Reddis* for construction, furniture, implements and fibre.

Forest Plants for Making Agricultural and other Implements

Most common implements seen with the *Konda Reddis* are *nagali, kathi, kodavali, goddali, golli, vaaluva, gonapam,* bows and arrows (*inti badha*) digging stick etc. All these implements are manufactured by them and approach the blacksmiths for iron parts. Every individual prepares his own plough with the help of the blacksmith. They use different kinds of wood for manufacturing different implements. The wood of all these trees is resistant to insects and last longer.

To manufacture this implement they use different kinds of wood for different parts. The wood of neradi, agise, bella, mokkeni, vepa, gummudu and chidugu trees are used to manufacture the main part of the nagali. They use these woods because of their strength. *Kadi,* the horizontal bar, which is tied to the neck of the oxen, is made of *gummadi* or *marri* wood, because it has a cooling effect on the oxen, and they are also lightweight and strong. *Medi,* which is vertical stick of the plough, is manufactured by the wood of *Bella, vandanam, korimi, neredu* and *rella,* which are lightweight and strong. The timber of *rella, korimi, vandanam, karaka, sunera, and tangedu* is used to manufacture *Veta,* which supports the plough, as those are also light in weight and strong. The rope for the oxen is prepared by *thati Nara* and *adda thadapa,* which they call *mokku thadu. Palupulu* is the rope prepared by two single ropes. *Varanam* is the rope, which they tie to the neck of the oxen. It is also made up of *thati thadapa. Kanne,* the rope which is inserted from the nose of the ox to the neck, is prepared by *adda thadapa.* The iron heads of the plough are also prepared by the Reddis with the help of *Kamara* (black smith).

Konda Reddis use bamboos for making bows and arrows. The bow is made up of bamboo while the string of the bow is made up of *jeelugu narra,* which is very strong. The edges are tied with *jeelugu naara,* which is a thin peel of the *jeelugu* tree. The arrow is made by a thin bamboo, which is fixed with an arrowhead made up of iron.

Bamboo is not only used for house construction and implements making purpose, it is also a source of food for the *Reddis.* Bamboo shoots are used to prepare tasty curry. *Reddis* also speak about edible bamboo rice (*Veduru atukulu*) that is rarely available in the forest. The bamboo rice is found on the end of the branches of 15-20 years matured bamboo trees. *Reddis* collect this rice by shaking the trees. They boil the rice and consume with curry as they do with the ordinary rice. Since it is rarely available, people do not rely on this but they enjoy collecting and consuming it when available.

Forest Palnts for Manufacturing Bullock Cart

Konda Reddis design and build their own bullock carts. This cart is used to transport the crop from *podupaka* to the village. It is made entirely of wood. Their indigenous knowledge of the qualities of different types of wood of local trees available helps them in successfully designing the cart. The wheels of the cart are made up of *vandanam karra* and *vegara* trees. The logs of wood supporting the wheels of the cart are made up of *nalla maddi,* and *bandaru* trees. The frame to be kept on the necks of the oxen is made of *gummidi karra.* The place where they sit on the cart is made of woods of *billa, agesi, andugu* etc. Rests of the things are made by *gesari karra, vandanam karra.* The reason for Konda Reddis preferring these particular trees for making the cart is their strength and their ability to withstand wear and tear.

Konda Reddis call wheels as *bandi pojalam; polligarra* is a vertical stick of the cart. *Bandi kadi* means horizontal stick, which connects the cart, and the stick, which rests on the neck of the oxen. *Bandi chirta* is the vertical stick on the cart, which is a support for the Konda Reddis to hold while sitting on the cart. *Bandi sajjuva* are the flat base on which things can be carried. *Taadi karralu* is a stick, to which a rope is tied to control the oxen.

Likewise, wood of trees like *bandaru, gummalu, doodugu,* are preferred for making doors, stools, and ground level stools. *Rokali* grinding stick is made from *peddapala chettu. Balli chettu* is used for making baskets called *errodi butta.*

Konda Reddis depend totally on the forests for fuel wood used for cooking as well as for lighting bonfires to fight the cold during winters. *Manta pullalu/karralu* are the terms that refer to firewood. They do not cut any trees for the firewood unless there is scarcity. They only collect fallen dry tree branches (*pullalu*) and dried branches of the trees. If they come across any dry and dead trees in the forest, they cut them and use as firewood. Plants used for the purpose of firewood are presented in the Appendix 15, followed by the local name and the name in English.

On certain days of the season (winter and summer), the *Reddi* women walk in groups to the forest and collect firewood for household consumption. They store the firewood close to the settlement and use it according to their requirements. In this way they make the best possible use of the available time and resources.

Konda Reddis also make charcoal from the wood of some specific trees. This charcoal is used for cooking in rainy seasons when it is difficult to find dried sticks and wood. This charcoal is strong and easily catches fire. They use the wood of trees like *Oddise, Neradu,* and *Veduru* to make charcoal. First they burn the wood and after it catches fire and spreads to the whole wood, they sprinkle water slowly and take out the charcoal. Reddis also use cattle dung (*peda*) for cooking purpose.

Blacksmiths use more charcoal to make iron implements like sickles, knives etc. Blacksmiths and *Reddis* use *tidurai* to sharpen their iron implements. *Tidurai* is a rough stone locally available on which iron implements like knives, sickles are rubbed so as to sharpen them nicely.

Treatment of Diseases by Means of Medicinal Herbs

The forest provides *Reddis* medicinal herbs. Folk practitioners use herbal medicines for almost all the diseases. There is a direct relationship between health and their environment. The majority of all forest plants collected by the *Reddis* are used for medicinal purposes. Other parts of plants are often used in other ways. In the present investigation I found that various plant species are used as medicine among the Reddis. The herbal medicines are prepared in various ways. For medicinal purpose, leaves, bark, fruits, seeds and root are often used. They are either ingested or rubbed on the body to treat one or more diseases or conditions. Most of the plant parts are used, by simply crushing a particular part of the species with a crushing stone and wood. But some are made by boiling the plant part with water. After decanting and cooling, the liquid is drunk or applied. Appendix 16 provides the information about diseases treated by means of medicinal plants collected from the forest by the Reddis.

In the village, *Reddis* go to medicinemen and women (*Vydudu/Vejju*) mainly to have their problem diagnosed, even though they know something about medicinal plants themselves. The effectiveness of the herb depends on knowledge of the exact nature of the disease. The response to a particular dosage may vary from person to person, while the same person may also react differently at different times. Thus, the patient himself will often experiment with a particular remedy and work out a suitable dose.

Medicinemen also believe that certain plant parts should be harvested during a particular stage in their growth. Since few people have detailed knowledge of the best stage or time to collect these herbs, they are almost totally dependent on medicinemen and women. The rules governing the time of collection, the part of the plant to use, storage, and preparation are known only to local faith healers, and for this reason they command considerable respect in Reddis society.

Reddis mostly depend and believe in herbal treatment and they also have a greater understanding of herbal plants. Not only the medicinemen, but also every elder in the village has a good knowledge of the medicinal value of some herbs, usually those plants used to treat common diseases like cold, cough, fever, stomach ache, swelling cuts, and wounds. While the youth of the community have little knowledge of herbs, they do have faith in the efficacy of these medicines. In general, members of these tribal groups tend to keep more detailed information about medicinal plants to themselves, and sometimes, medicinemen do not even pass on their knowledge to members of their own family.

It is clear from the above description that the traditional knowledge system of the medicine to cure various diseases among Reddis depends upon their forest environment. Reddis still use herbal medicine even though modern medicare is available to them. Due to lack of modern health facilities, readily available they depend on herbal medicine.

With the passing of each generation, the knowledge of medicinal plants is being lost, despite the continuning importance and relvance of that knowledge to the Reddis. This knowledge forms the basis of health care in the community. Even if western medicine acquires dominance over traditional medicine, its high cost, compounded by an increasingly tighter economic situtation in the community, will continue to force the Reddis to opt for easily available and cheaper local solutions. Medicinal plants are readily available in the community, while advice and consultations can be paid for through non-economic reciprocal exchange.

A major cause for the loss of indigenous knowledge among Reddis is the disruption of traditional channels of oral communication. This has made it difficult for older generations to pass their knowledge on to the younger ones. Neither children nor adults spend as much time in the community as they used to even a generation or so ago. Most children aged 6-12 years now attend rural schools, where indigenous knowledge rarely forms an intergral part of the curriculum. On the contrary, formal teaching in the community tends to be guided by a philosophy of education. In the process, *Reddi* children are subtly and sometimes not so subtly indoctrinated with negative attitudes towards anything that is 'indigenous'. As a result when they become adults, some of them may not value their cultural knowledge as highly as their parents or other elders before them.

The oral tranference of indigenous knowledge is also hindered by urban migration. Many Reddis in the region travel to the city on a daily basis to look for work or to sell farm produce. Some are forced to migrate for much longer periods to cities further away. And finally, the loss of ethnobotanical knowledge is directly linked to the loss of biodiversity. Some *Reddis* remarked during the interviews that certain medicinal plants were becoming more difficult to find, necessitating trips to other communities or regions.

Detailed study on the ethnomedicinal practice of the *Konda Reddis* has not yet been made. Though the *Konda Reddis* have been using these medicines for quite a long time, it is very difficult to judge the effectiveness of the traditional medicines. However, further research on these traditional medicines will provide an insight to improve the quality and effectiviness of the same.

Economic Dependency of Wild Flora

Since forests surround Reddi dwellings, they are wholly or partly dependent on the forest for consumption as well as commercial needs. There is seasonal variation in the availability of different forest products. Men, women and children are engaged in activities related to the gathering of minor forest produce to supplement their meagre income. Though marketing network is not well developed, *Konda Reddis* sell forest produce to Girijan Cooperative Corporation (G.C.C), local traders (non-tibals) or other petty shops. Appendix 17 discusses the commercial uses of various wild plants products identified and used by the *Reddis*.

Previously, nearly 40 items of Non Timber Forest Produce (NTFP) were being gathered by the tribals and sold to G.C.C. Over the years, large-scale forest degradation has placed considerable pressure on the community and seriously affected their livelihood. This deforestation has had effect on the domestic needs of the community and on NTFP, which are part of the traditional household economy. Deforestation has had effect on G.C.C also because at present there may be not more than 25 items of NTFP being gathered by tribals and sold to G.C.C. The G.C.C. has been affected by various factors like intervention of private businessmen in the tribal areas, lack of reasonable price and transport system in interior tribal areas, migration of youth from tribal areas to plains in search of employment. Due to the Government intervention through the Integrated Tribal Development Agency (ITDA), Girijan Cooperative Corporation (GCC) and the forest department, developmental changes have taken place. But again they have to make planned efforts to improve the quality and quantum of production. They have to help the tribals in identifying value added forest products. They also have to make efforts for regenerating select plant species that have medicinal and commercial value.

Konda Reddis' subsistence economy shows that they are totally dependent on their immediate forest ecosystem. Any change in their ecosystem is going to affect their lives immensely. Due to their illiteracy and innocence they are largely exploited by the non-tribal people. Due to lack of transportation, they are not able to sell their forest produce in the market. Instead the people from plains come to their hamlets and barter their forest produce for broken rice or any other edible items. These intermediaries exploit them to the fullest.

Therefore, planned efforts are necessary to usher integrated development and provide necessary infrasture to reach remote areas of *Konda Reddis* by developing the necessary transportation and communication facilities. The government could establish a market and the information regarding the prices for various commodities must be provided at all the market centres organized in the region for developing trade in farm and non-farm products. Roads and other communication facilities are to be well developed for the overall development of the economy of the *Konda Reddis*.

Rajulba Panduga

This is the festival of forest God *Rajulba Panduga* that is celebrated every year by men only at the end of Agricultural activity. On this festival at least one person from each house should celebrate in their fields by sacrificing chicken in front of *Putta* (anthill). Because, after harvesting the crop, first they make an offering to the god and then only do they eat. They believe that this ritual helps the villagers in judging the outcome of the crop to sow in the coming season. It is performed to protect the crops from the evil eyes, evil spirits and wild animals from the forest. According to them, after sacrificing, they cook the chicken in their houses and eat together. This festival is celebrated in the month of February.

Foraging Technology

They use simple tools to forage the forest products. They use five types of knives for this purpose i.e., *geetha kathi, mattakathi, potakathi, chethikathi, vankakathi* and use digging stick or *valva* and bamboo basket or *budagampa. Geethakathi* is used for tearing fruits, leaves; bamboo etc. *matta kathi* is used for cutting, bunches of fruit. *Vanka kathi* is used for cutting long and big parts of tree like *taati aaku*. The *chelupu* or *chethi kathi* is used to clear the bushes, creepers and also used to collect leafy vegetables, mushrooms etc. *pedda kathi* or *potakathi* is meant for the collection of edible fruits and vegetables available in the forest. The digging stick is another instrument used to collect different varieties of tubers and roots. The implements used to dig out the tubers are *gunapam, gobbam, thatta, Gunapam* is iron crowbar, while *Gobbam* is small iron instrument and *Thatta* is a basket, which is made of bamboo used to carry the tubers.

Inorder to carry the collected forest produce they use the bamboo basket, which is exclusively meant for the purpose. There are two types of axes i.e., pedda goddali and chinna goddali or chila goddali. Sickle *(kodavali)* is used exclusively at the time of harvest to cut the corns. Small digging stick *valva and* large digging stick *gacchala* are used to dig the holes at the time of sowing and also at the time of weeding to pick out weeds, creepers and bushes. Both digging sticks are multipurpose in use and are used to collect tubers, roots and other minor forest produce. *Cheta* is made of bamboos and is used for winnowing. *Chelupu kathi* is used to cut the branches and to clear the undergrowth. Sickle (kodavali) small digging stick valva, large digging stick, gacchala is an instrument used for weaving *cheta* (basket). *Chelupu kathi* is used to cut the branches and to clear the undergrowth. The *goddali* is an important instrument, which is made into an in finished crude blade and is inserted into a conical shaped wooden piece. It is used to cut and break the trees and wooden pieces.

The knowledge of plants is perhaps the most refined aspect of local knowledge among *Reddis*. In almost all cases, classification and nomenclature of plants are based on plant morphology and use. In many cases, the same name may be given to different plants (a "generic" name), usually because there is no practical reason for differentiating the plants, although the people do know the differences. Finally, the same plant may be classified and named according to different criteria, resulting in several names for the same plant. Usually the local people's naming, knowledge and classification of plants are more pragmatic and immediately utilitarian than the formal scientist's. The local people also have an intimate knowledge of the characteristics and values of different plants. Therefore, the value of plants is a relative one that varies with the environment, time and cultural background.

This chapter has examined the *Reddis*' local knowledge in a number of forest plant species used in large quantities within the village ecosystem. *Reddis* possess local knowledge about the utility of numerous plant and animal species available in the forest. Their knowledge about flora and fauna of the area in which they inhabit is extraordinary and they have devised several techniques for their utilization in a sustainable way. The *Reddis* know exactly which type of products to collect, as well as the proper season and time of day. The tools and implements devised by them are very simple. The various food procuring devices like bows, varieties of arrows, traps, snares etc indicate that primitive tribes are not wanting in ingenuity and skill. They never overexploit the forest resources, and are very precise about the time that a particular plant or shrub should be picked, dug or cut.

In the above, an attempt has been made to present extensive facts and figures on the local knowledge use of forest plant products among the *Konda Reddis* of Andhra Pradesh. That local knowledge is closely bound with their way of life. This has faced a series of challenge in the recent past. The rules and regulations that have traditionally governed them are breaking down and the homogeneity of the individual society is giving way to heterogeneity of settlement. The traditional beliefs and indigenous knowledge of the people are now being questioned by the younger generation, whose task is to keep that culture alive. The causes of these problems are many and have to do with economic, social and political factors. The traditional role of neighbours acting as each other's 'keepers' during misfortune or happiness is diminishing.

It is, therefore, clearly understood how *Konda Reddis* are closely dependent on their forest resources for their survival. From such natural resources they not only draw the required energy for survival but also get sufficient protein, fat and minerals, which are essential for existence.

During the fieldwork, it was observed that their forest and agricultural produce cost high in the market. The produce is brought from the tribals for a very low price or for other provisions. Some times *Reddis* do exchange their minor forest produce at shandies/petty shops for bare necessities like salt, oil, legumes, beedies, sugar, etc.

The younger generations coming in contact with the non-tribals are losing their cultural values and faith in their beliefs systems. It was found that due to the intrusion of outsiders and stringent forest laws, the feeling of possession of forest as "our forest" among the tribals who used to protect and safeguard it as their own, is lost. Due to poverty, scarcity of forest produce and difficulty in procuring these resources, the younger generations are misguided and are lured by easy money. The old men from the tribe reported that the younger men are lured with cash for cutting the trees in the forest.

They are paid a lump sum for cutting trees overnight, which are towed away by the non-tribals later. It was reported that the deforestation is not only by these tribal men, but by the non-tribal and even the forest officials who employ people to cut trees and supply to their seniors in order to appease them. It was reported that there has been a connivance of the forest officials and non-tribals in the illegal cutting of trees, which mainly occurs at night. During the fieldwork I found many trees had been cut down on the way to *Konda Reddi* hamlets.

As a consequence of excessive felling of the trees by vested interests, *Konda Reddis* have to walk longer distances in search of minor forest produce and firewood. The strain on these *Reddis* is further added due to the non-availability of forest food and decrease in the food grain production. Further, the availability of forest food is reduced as a result of deforestation and lower fertility of the soil. Thus, it can be found that environmental degradation has detrimental effect on the daily lives of the *Konda Reddis.*

Konda Reddis' complete dependence on the immediate forest ecosystem proves their age-old adaptation to their environment. It is in recent years that they are now being exposed to outside world and their adaptability is being challenged. Due to the interference of outside non-tribals and forest policies and programmes they are being exploited, alienated and losing control over their environment and thereby struggling to survive. The loss of forests poses great problems especially to eke out subsistence. They are now struggling much more to collect firewood, fodder and minor forest produce. Further, *Reddis* are losing command over their resources on account of pressure from the non-tribals.

Another factor responsible for change is the migration of youths from tribal areas to the plains and semi-urban and urban centres is to take up employment there. In the course of time these youth begin to relinquish their traditions and adopt modern culture. Yet another problem in maintaining traditional culture and indigenous knowledge results from the loss of natural forests. Forest depletion caused by various factors has led to the loss of natural pockets of indigenous flora and fauna. With the degradation of forest, they do not get any kind of fruits in good quantity. The need for documentation is doubly acute in view of the rapid socio-economic and cultural changes taking place in this tribal community.

5

Local Knowledge and the Use of Wild Fauna

Indigenous communities inhabit varied geographic and climatic zones of their respective regions. Forests have been their dear home and their relationship with forest is symbiotic in nature. They utilize the resources without disturbing the direct balance of the ecosystem. Living close to nature, the indigenous communities have acquired knowledge about the use of wild flora and fauna most of which is not known to the outside world. This rich knowledge system if subjected to scientific scrutiny could benefit humankind in many ways. The inroads of modernization are presently posing imminent danger for this rich and varied expertise and wisdom of ages and it is likely that this will be lost to humanity for all time to come. Habitats where indigenous communities lived and the environment in which folklore evolved on the uses of wild plants and animals are fast disappearing.

The relationship between humans and animals is an age-old phenomenon. Humans have been using animals for various purposes such as food, clothing and medicine from prehistoric times. But in recent years, due to large-scale urbanization and indiscriminate cutting down of forests, the very existence of the treasure of indigenous knowledge on animals is threatened. Besides, scarcity of written documents and lack of proper communication between the indigenous and urban communities add to the problem.

This chapter aims at describing the identification, characterization and description of some of the faunal resources and their utilization by the Konda Reddis. The present data is based on information gathered through interviews and group discussion with the village elders. The village selected for information is in an interior place where the local beliefs and indigenous practices are performed and where people have the knowledge of identifying the wildlife and their traditional use in their community. This chapter accounts for the Konda Reddis knowledge of wildlife in their own habitat and hence, gives a glimpse of their rich traditional environmental knowledge.

Reddis use wildlife resources in many different ways reflecting the local, economic, cultural, social and ecological differences of this vast land area and its many varied climates and socio-economic systems. It is accordingly somewhat arbitrary to classify the pattern of resource utilization. But it is useful in terms of organizing available information. Each individual species is intrinsically valuable for its own unique features. Thus, people acquiring additional socio-economic value as a resource, directly utilize some species.

Classifying Patterns of Utilization

Konda Reddis classify animals into two categories i.e. domestic animals (*Pempudu jantuvulu*) and wild animals (*Adavi jantuvulu*). The animals domesticated by the *Reddis* are oxen (*Yeddulu*), cows or *Bos indicus* (Avulu), calfs (*Dudalu*), young oxen (*Geethalu)*, goats (*Mekalu*), sheep (*Gorre*), buffalow (*Gedde*), bulls (*Dunnapothu*), poultry (*Kodlu*), dogs (*Kukkalu*), cat (*Pilli*), etc. They classify wild animals as *Jantuvulu* and sub divide them into birds (*Pakshulu*) snakes (*Pamulu*), insects (*Purugulu),* flies (egalu), fishes (*chapalu*), Chimalu *(ant)* other wild animals (*Adavi jantuvulu*). *Konda Reddis* easily identify various animals found in the forests.

Konda Reddis classify wildlife according to the species (*Jati*), their usefulness, sex and size of the animal, height and weight of the animal, a distinct character of the animal, different body parts of the animal, colour of the animal, and also whether or not the meat of the animal is fit for consumption. Further the chapter includes discussions on inventory of wildlife, their general characteristics and behaviours, hideouts, their diseases and treatment, use of wildlife parts as medicine, native concept of reproduction cycle of wild animals, hunting practices of wild animals etc.

Konda Reddis categorize the wild animals into different species by various criteria, such as sex, age, size, hiding place, whether or not their meat could be eaten, and also by some distinct behavioural characters and features of the animals. For instance, they categorize wild cats (*Adavipilli*) into different types according to their colour, features, and general characters, and suitability of their flesh for food. Wild cats are classified into *Kurudepilli, Bothapilli, Manupilli, Thompilli, Battugupilli, Bourupilli, Chevulapilli, Mungipilli*, etc. *Pilli* is the local term for the cats. Similarly *Konda Reddis* categorize rats into two types according to their hideouts: rats found in human residential areas and wild rats. Rats, which live in their houses and villages are again of many types, such as *Sunchieluka, Gaddeeluka Surieluka*, etc. and wild rats, which live in the forest and fields, are classified into different types according to their colour, size, features and other characters. The wild rats are differently classed as *Rassieluka, Chettieluka, Uduthaeluka, Vedurueluka, Chemalu mingueluka, Elutaeluka* and so on.

Size of the wild animal is always an important criterion for categorization. For instance, they categorize leopard (*Chirutha puli*) on the basis of their size into 2 types: *Akuchidu* (which is smaller in size) and *Bandichindugu* (which bigger in size). Like wise a *Kanasu* is categorized into 3 types: those that are bigger in size (*Nallakanasu*), those medium in size (*Errakanasu*) and those smaller in size (*Gottekanasu*). The *Valugu* are categorized into 3 types based on the size of the species. *Maddivalugu* are big in size, *Kondavalugu* are medium in size and *Kodivalugu* is smaller in size. The varanus monitor (*Udumu*) is also categorized into 3 types based on its size as *Langaudumu*, which is big in size, *Kondaudumu* is medium in size, and *Galiudumu* is smaller in size. Similarly the wild sheep (*Gorre*) is also categorized based on its size: *Adavigorre* is big in size, *Dabbagorre* is medium in size and *Ethagorre* is small in size. The fox (*Nakka*) is categorized into 2 types: *Soulnakka* is bigger in size and *Konkanakka* is smaller in size. *Reddis* also categorize the animals according to their sex: a male is called as *Pothu* and a female is called *Adadi* or *Pedi* irrespective of the species to which it belongs.

General Characteristics and Behaviour of Wild Animals

Mammals

Wild Buffalo or Bison (Gorra gedde)

Konda Reddis identify bison as a large wild fauna, similar to cow but having a lager head and hairy shoulders. At the lower part of the legs from the knee to hoofs and a part of the neck is white in colour. Buffalo horns are round in shape; light red in colour, and the whole body is covered with long hair. The animal has a long tail and the end of the tail has a white colour patch. The periodicity and frequency of reproduction in the forest is apparently unknown to the Reddis. Male is called *Pothu gedde* and it has white spot on the forehead. Female is called *Pedi*. The male buffalo has big horns and the female has no horns. Both male and female can run fast in spite of their bulky bodies. The male is very aggressive. It chases human beings if it sees any, whereas the female is less aggressive and usually stays away from the humans.

According to the Reddis wild buffalo feed primarily on tender bamboo shoots (*Pidem*), tender leaves of bamboo, leaves and bark of *Kinneri, Gotte kayalu, Gummudu kayalu,* and *Soudu* (it is a salty soil, which the animal eats to digest the food). The animal eats *Soudu* before and after food. They may also eat berries and other bland foods of plant origin. Apart from the forest produce the animal sometimes stray into the fields and eats the crops like horse gram (*Uluvalu*), red gram *(Kandulu)*, green gram *(pessarlu)*, black gram (*Minumulu*), gourd (*Anapa kayalu)*, brinjal (*Vanga)*, cotton fruits and leaves *(Patti kayalu and Akulu)*.

Wild buffalo generally moves during the day and the night for food. It eats the leaves of tree by bending the branches of the tree. It moves to the fields to eat crops and spoils it completely. The animal moves to canal to drink water in the evening from 6:30 to 7:30 P.M, and again in the early hours from 3:30 to 4:30 A.M. It likes to take rest and sleeps in *Bandam* place (a plain place of land in the interior forest in rainy and winter seasons). And in summer season, it sleeps in cool and wet places and also rolls in the mud to cool its body.

The mating season of wild buffalo coincides with the onset of the rains in May. The gestation period is 280-300 days i.e. ten months. The animal calves being born in June or July. Calves reach sexual maturity in two to three years. After attaining maturity, it starts to wander alone. At this stage, it has the capacity to give birth to babies till the age of 15-16 years. After 15-16 years it becomes weak and old.

The bison can be attacked by the disease called *Gadlu* (boils) near the hooves. To get relief from the boils, the animal walks in a muddy place and also eats some wild bark to be cured. The dung of the animal looks like buffalo dung. Reddis identify the animal by seeing its dung. Reddis weigh the animal locally by *Jeddi* and the wild buffalo usually weighs 4 *Jeddi* (about 400 kgs). If they hunt, eight people are required to carry this animal. *Reddis* generally do not hunt this animal purposely because they do not eat the animal meat.

Deer or *Axis axis (Duppi):* Reddis identify this animal as light brown in colour with white dots on the whole body. The ecological range of the animal is quite unusual, it may be found in lightly wooded areas, brush land, thorn bush, and swampy areas. It is tolerant to altered environments and can thrive in successional mosaics and partly deforested or cropped land. Population densities can be high in some areas. Deer lives in small groups of 2 to 6 or more. The species may occupy relatively permanent small home ranges. Deer are active in the morning and afternoon but they may become strictly nocturnal and extremely shy when constantly pursued.

According to *Reddis* the deer eats tender grass (*letha gaddi*), leaves of *Mulla akulu, Parami pandlu, Regi pandlu, Chekkum pandlu, Korukudu pandlu, Usiripandlu, Shikai*, etc. It also eats *Soudu* for digestion of the food. But dietary selection varies greatly from one place to another. The seeds and fruits of certain trees are a prominent item. It feeds on all agricultural crop varieties of pulses, cotton, brinjal, chilies maize, sorghum, beans etc. Hiding place of the animal is *Chitti adavi* (open places in thick vegetation), big rocks *(pedda rallu)*, water is the limiting factor for the animal.

The mating season is during summer and rainy seasons. The gestation period is 250-280 days i.e. nine months sometimes seven months, the young being born in June or July. Young deer reaches sexual maturity in one year and is ready to give birth. The female deer can give birth to the offspring till it attains an age of 12 years, after which it becomes weak and old.

Reddis call male deer as *Pothu* or *Narugupothu* and female deer as *Pedi*. Male deer has antlers (wide branch like horns) while the female do not have horns. According to Reddis the male deer shed their horns every five years, which are fully matured. Babies are locally called as *Kunalu*. Small size deer is named *Gotte*. The excreta of the animal looks like that of goat. It can be attacked by the disease called *Gadllu* near the hooves. The weight of the animal is approximately 100 kgs that is one *Jeddi*, in local measure. If it is hunted it has to be carried by two persons.

Cervus unicolor (Kanasu): This is a strictly forest-dwelling species, preferring closed, moist, lowland or montane forest, but may also frequent cultivated areas and clearings at night. It is found in a great variety of habitats from savannah, forest mosaic and in well-developed deciduous forests. Judging by the importance of this species in subsistence hunting, it must be relatively abundant over much of the forest.

Kanasu live alone or in pairs and are apparently sedentary. Visibility is not very poor in the habitats of these shy, or nocturnal animals. Male have horns while the females do not have horns. Male animals sheds the horns after they mature once every five years. *Reddis* call the baby animal *Lagaduda*.

According to Reddis, this animal feeds on *Pidem* (leaves and tender shoots of bamboo). Food is same as that of *Duppi*. The animal eats *Soudu* before and after food. This animal sometimes strays into the fields, eats and damages crops like *Uluvalu, Kandulu, Pesarllu, Minumulu, Anapakaya,* Cotton leaves and fruits, Brinjal etc. This animal is nocturnal and generally starts hunting for food around 6:30 p.m. and sleeps after having food. The excreta of the animal looks like that of goat which *Reddis* locally call as *Goddelu*.

Mating season of *Kanasu* is timed to coincide with the onset of the rains in May. The gestation period is ten months (i.e. 280-300 days), and young animal is born in June or July. After three years the animal is ready to reproduce till 15-16 years. After 14-16 years, it becomes weak and old. After attaining maturity it starts to wander alone. When the animal gives birth to the baby, the baby, starts walking after five hours.

The animal can be afflicted by boils *Gadlu* (boils) on the limbs. It is the main animal used in subsistence-hunting in the forest areas. They are hunted with dogs. Whereas, other techniques used are the same as in deer hunting, which is mentioned above.

Wild Boar or Sus scrofa (Adavi pandi): According to the Reddis the animal is a large, fierce, hairy similar to a pig (*Urapandi*). The body of the wild boar is big, it has a long mouth and is covered with thick black hair. The skin of the animal is very strong and fatty. This animal is a nocturnal feeder and rarely comes out in the daytime. The animal lives in stable family groups usually made up of a mated pair and their cubs and sub adults may remain in the parental group up to the age of two. Male boar is called *Vontadum* and the female *Dadda*, while the piglets are called *Kunalu*. Males have the pair of small tusks coming out of the corners of the mouth and the female has none. The hiding place of the animal is dense bushes and big rocks. Boars sleep in a trench dug out with their snouts (*Goti*). In summer, to get relief from the heat, they roll in wet mud *(burada)* for cooling their bodies. The weight of the animal is 50-75 kgs. If the animal is hunted four people are required to carry it. The animal is very aggressive and attacks directly from the front on the right side of the person. The wounded animal is very dangerous. If this animal sees any person it directly attacks. The excreta of this animal is similar to that of horse, and it is locally named as *Goddelu*.

It is both herbivorous and carnivorous. The animal feeds on a great variety of crops like stem tubers (*Dumpalu*), roots of crops, rice *(Vari)*, Jowar *(Jonna) Gantte, Sollu, buduma, Tati tegalu, Tati pandlu,* mango kernel (*mamidi tenkalu)*, berries (*Neradi Pandlu*), tender bamboo shoots (*Veduru kommulu)* etc. The animal also eats insects like earthworms, *Laddepurugu*, and dead animals' meat. The animal digs the ground with its snout to take out the roots, tubers, earthworms, insects, etc. This animal always strays into the fields for food and damages the crops.

Mating season of the wild boar is between November and December and the gestation period is seven months. The piglets can be seen in the months of June to August. The animal gives birth to 10-15 piglets once in a year. The animal when delivering the piglets for the first time gives birth to 4-6 piglets, the second time 6-10 and the third time upto 10-15 piglets. Piglets attain sexual maturity in one or two years and the animal is ready to reproduce babies up to 15 years. Wild boars are mainly hunted at night with torch, and often with dogs. Other hunting methods include pulling them out of their dens, trapping them, or simply waiting along one of their trails. The meat of the animal is very tasty.

To protect the crops from the wild boar, Reddis tie *Uchu* with a clutch wire at corners of the field on the way of the wild boar. Other methods include setting of country bombs (*Natubombulu)* on the way or to construct some traps like *Kuddelu* and *Inti badda mette* to trap the animal. If the bomb is crushed with the mouth, the animal dies on the spot. They buy bombs from the local market to kill the animal. The traps are made by stones and dry wooden sticks; if the animal enters the trap, all the stones fall on it and it cannot escape from the trap.

Wild sheep (Konda gorre): This is a strictly forest-dwelling species, preferring closed, moist, lowland forest, water bodies, but may also frequent cultivated areas and clearings at night. The animal looks like a domestic sheep and the average height is about three feet. Males (*Pothu*) have horns with 3 branches and the females (*Pedi*) have none. They live alone or in pairs and are active in the morning and afternoon. According to the Reddis, the animal is commonly found in small open places in the interior dense forest (*Chitiadavi*) and also in the *Tutlu* (dense tendril places). This animal sleeps in the night on plain ground in the interior forest. The peculiar habit of the animal is that it defecates in the same area where it sleeps. This animal is one of the fastest runners in the forest and it generally shy and keeps away from humans. Whenever it goes for food it comes back to the same place for sleeping. If it cannot find its place, it is upset and makes noise.

These animals mostly go around in herds for grazing. This animal cannot survive for more than 24 hours without water. These sheep go to nearby streams or canals for drinking water early in the morning and again in the early evening.

Mating season of the animal is in summer. Gestation period lasts for about 7 months. The fawns (baby animal) are born in April-June. Fawns reach sexual maturity at approximately one year. Reproductive capacity is low with an average annual rate of one fawn per year up to 10-12 years. Fawn begins walking within two hours after birth.

This animal is very much hunted in the forest. They are hunted with dogs. A frequent technique is to drive the prey down to a river or stream where the hunters wait. Nocturnal hunting is done by using torchlight on the animal. Meat is tasty and delicious and usually eaten fresh or sometimes, dried and salted. It is a preferred and is an important item in the community diets. If the animal is hunted down, needs two persons to carry the animal.

Porcupine or Hyxtris indica (Mulla pandi or Nedi pandi): This has a protective covering of long sharp quills and is found in great variety of habitats from forest to bushes. The animal belongs to the class mammalia, order rodentia. It is a nocturnal animal usually live in burrows, rocky areas and caves, locally called *Surgan*. The special character of the animal is that it has spines (stiff needle like hairs on its back) all over the body. The sharp spines of the animal are about six inches long and are used for protecting the animal from enemies. These quills are periodically shed when it is fully matured. It has a sharp mouth with which it can dig roots and also can make holes easily. According to Reddis the limbs of the animal resemble the limbs of the dogs. The average weight of the animal is up to 10-12 kgs. Porcupine is very active and a very fast runner. Mostly these animals go about in a group for food. Group size varies by season, habitat conditions and population density.

This animal feeds on a great variety of tree fruits, seeds, bark, roots, leaves and different crops and eats bulbs, fallen leaves, flowers, medicinal plants etc., It is quite selective and the main food in its diet may vary from one place to the next. Porcupines look for roots (*Verulu*). When fruits are scarce they live on shoots and leaves, sometimes helping themselves to various crops like tubers (*Dumapalu*), pulses and vegetables. They are very fond of fruits, mango, mango kernel etc. It is said that they are very fond of mangoes. The excreta of the animal is similar to that of the goat but the solid waste shape is slightly different. The mating season of the animal is summer. The gestation period is 5 months and piglets (baby animal) are born in the rainy season.

Porcupine is in the list of subsistence hunting game. They are also hunted using dogs, which drive them into burrows or hiding areas. And the usual technique now is to set traps along pathways. Trapping becomes the main activity for Reddis during the hunting season. Of all wild and domestic animals in the community porcupine meat is considered the most exquisite. It is nutritionally and medicinally very important among the Reddi communities and there is also a great demand in urban areas. The meat is used as a medicine for unhealthy people. Reddis say that the meat of the porcupine is very tasty and any ill person can eat the meat of this animal because this animal feeds on medicinal plants also.

Rabbit or Wild hare or Lepus nigricollis (Chevula pilli or Kokkupilli): This is a small animal with long ears and large front teeth, which moves by jumping on its long back legs. Rabbits are game animals and usually an adult is approximately 49-69 cm long and weighs 3-5 kgs. But the females tend to be a bit bigger and heavier. The female hare *(Adadi)* is stronger than male *(Pothu)*. Hiding place of the animal is *Tupallu* (thorn bushes or small bushes) to open deciduous forests, scrub forest (*Kampalu),* secondary stubble vegetation and cultivated areas.

This animal feeds on green grass like *Garki gaddi,* pulses, tender brinjal leaves and fruits, *Soud* etc. The peak time for searching food is around 4:30 to 6:30p.m. the rabbit sleeps regularly in the same spot if the situation is bad, only then does it change its place.

The average gestation period is 1-2 months. The average production per female is probably 2-4 per year. The average production of babies continues till four years; after five years it becomes old.

A peculiar habit of this animal is that it can move without making any noise. Wild hare is one of the fast runners among the animals, and shies away from the people. It is very difficult to catch or hunt the animal. It sleeps in burrows and changes the place if it senses any danger. This animal always strays into the fields and damages the crops and hence to protect the crops Reddis fix traps in their fields.

The most effective and probably the most destructive method is lighting the bushes at night with a torch. A more sporting technique is to hunt rabbit by day with dogs or to simply search for rabbits in the bushes. Reddis hunt rabbits by setting nooses or traps along rabbit pathways in the fields or use slingshots or lure them by imitating their calls. The white and tasty meat of the rabbit, the only material product, is widely eaten by the *Reddis* in local area where the rabbit is most abundant and sometimes the only game animal. Daytime hunting of rabbit also has great recreational value.

Bear or Melursus ursinus (Elugubanti or Elugubuddi): This is a large strong wild mammal with thick black furry coat that lives specially in cool and wet area of the forest. The colour of the animal is pure black. It has two long and sharp white nails on the finger of the limbs. Reddis categorize this animal into three types according to the size and age of the animal. *Puruelugubanti* is a small sized animal, which is about 2-3 foot tall and the weight of bear is around 25-30 kgs. *Banti* is the medium sized animal, which is about 3-4 foot tall and weights around 45-50 kgs. *Dasaribanti* is a big sized bear, which is about 5-6 foot tall and it stands on its hind limbs. *Dasaribanti* has white mark on the forehead that indicates the old age of the animal and weights is around 95-100 kgs.

Reddis call the male bear *Pothu* and female *Adadi*. Female is very much knowledgeable in taking care of the babies. If a person is seen coming towards her, she climbs the tree and protects the cubs. All bears have human features like mammary glands to feed the young ones. Footprint of the animal is similar to man but slighter in length and width. Hiding place of the animal is on rocks, caves and big size trees in the interior forest. When the animal is wounded and is unable to go to its place of resting, it hides in thick bushes. *Dasari Banti* is very strong; it goes for food in the night and sleeps in daytime. It has a sound sleep and if any body goes and makes noise in front of the cave it cannot listen or will not come out of the cave. Though this animal is aggressive, it usually shies away from people, but it attacks when it feels threatened or in very close encounters with people. Reddis say if the animal sees any person from close quarters it directly attacks the eyes of the person. If observed from long distance it will not try to attack and simply goes on its way. Else, if it understands that the person is coming towards the same direction it attacks on knee and elbow of the victims to make them immobile or otherwise it will hit the testicle (scrotum) of the person.

It is both carnivorous and herbivorous; fruits, insects and leaves are the basic items as food of the animal; it feeds on insects like *Chemalu, Chedapurugulu,* ants and other insects, which are available in the forest. Bears are great fruit eaters and they eat fruits like jack fruit (*Panasa*), *Gidipandlu, Rella, Pardonda, Nallagidi,* Soudu etc. In summer season the animals go for food all around the forest. Bears are more active in summer because they get to eat a variety of fruits. In winter season they move in groups for food. The general behaviour of the animal in winter season is to move over large areas because in winter the fruits are not available in the forest. It can climb on the tree and eat fruits and is quite fond of honey. The movement of this animal is identified by Konda Reddis by its footprints.

Sexual maturity probably occurs between the age of five and eight years. Mating usually takes place between March and May months. These animals are more active in summer for mating because they get to eat a variety of fruits. At the time of mating the animal makes noise in the nights. During the mating season, females are active and they make mating calls and attract the males for copulation. In mating seasons, the females separate and go in search of partners. The Gestation period among these animals is about 9-10 months. A female bear gives birth to 2-4 cubs. The cubs can be seen in the months of September to January. The cubs require maternal care for many months and may remain in the mother's territory as long as two years before moving on independently. After delivery, the female takes care of its genitals and keeps them dry, as it leads to infections if wet. A female gives birth to young ones till it is of 30 years. The female bear get genital diseases in winter season due to contamination of water.

The fat of the animal is used as medicine for curing joint pains (*Keedlanoppi*) and flesh used for waist pain, fits (*Morsha)* and during delivery. Generally they take out the fat part from the body of animal. After taking fat it is heated for half an hour and slowly it starts melting and forms oil. This oil is stored in a container for long time and they used it as a medicine whenever they like.

Leopard or Panthera pardus (Chirutha pul or chindugu): It is a large wild animal that has yellow fur with black spots on it and lives in tangled moist forest along rivers, swamp and is also found in deciduous forest and mountainous (*Konda)* zones. The *chindugu* is primarily a nocturnal, solitary and sedentary animal. It marks its territory by loud roaring, by urinating and by scratch marks on trunks. It may travel kilometers every night or may remain two or three days near a recent kill. During the hot hours of the day the animal likes to rest in thick vegetation or occasionally along branches up a tree. Generally the hiding place of the animal is in dense forest, big trees and rocks etc. Its size depends on sexual dimorphism. The total length of adult males varies from 160-225 cm and the weight from 64-114 kgs and females may be 140-185 cm and 45-82 kgs.

This animal is a carnivore and feeds on a great variety of vertebrates, rodents, rabbits, deer, large birds, rats, monkey *(Bandalu), Soudu* snakes, water turtles and tortoises and their eggs, fish and may prey upon domestic animals like sheep, goats, cow, small pigs and decayed parts of plants. They kill the larger ones by leaping on them and inflicting a mortal bite on the neck. The prey is then dragged off to a tree-shaded place.

Animals reach sexual maturity in three years and they apparently mate once a year particularly in the summer season. Gestation period of the animal is ten months. It is said that a female animal can reproduce till it is fourteen years of age. A female leopard gives birth to 2-3 cubs in a litter and the cubs are born in winter. Cubs are cared for by mother for several months and they remain in her territory up to the age of two. The female animal hunts and feeds the cubs. Reddis categorize this animal into two types according to their size. Smaller one is called *Chindugu* and bigger one is called *Bandi chindugu.* People distinguish chirutapuli by gender, where male is called *pothu* and female is called *Pedi.*

Both adult male and female hunt for prey where as small immature ones do not go to hunt. Females go for hunting after giving birth to the cubs for their food. Konda Reddis identify the movements of the animal from its footprints, which resemble a dog's footprints but bigger in size.

Monkey (Kothulu): They are mainly lowland animals but some species can be found in forests. Monkeys require wooded habitats and some monkeys prefer deciduous forests. According to *Reddis* the animal has a long tail and climbs trees. Its lips and buttocks are red in colour. Monkeys vary in size. The weight of an adult animal is approximately up to 10-15 kgs.

As a group monkeys are tree dwelling, sedentary and social in habit. Some are noisy and easy to detect, whereas others tend to flee or are mostly silent. A few live in small family groups made up of a permanent pair and its offspring of various ages. These animals move round in groups. Some customarily live in bigger groups ranging from 5 to 20 or 30 individuals, in a more complicated social structure and are usually territorial in habit.

Monkeys feed on all kinds of fruits like mango (*Mamidi*), tamarind (*Chintha*), seeds of cashew nut (*Jedi pickelu*), *Dosa kayalu, Gumadi, Turam* fruits and leaves, Tubers (*Saggu dumpalu*), Cotton fruits (*patti pandlu*), brijal (*Vanga*), *Berakayalu, Mokkajonna,* most vegetables and almost all food crops. Fruits, insects, leaves, and vegetables are the basic items in their diet. They are great fruit eaters but also invest long hours looking for food of animal origin, including insects, molluscs, small vertebrates, bird eggs, chicks, etc. It is said that monkeys eat less and waste more.

Monkeys generally reach sexual maturity in the second year of life but do not breed so long as they remain in the maternal group. Some monkeys take 4-8 years to reach sexual maturity. The gestation period ranges from 128-145 days. They almost always give birth to two young ones. The interval between successive births may be less than one year. Many monkeys apparently breed year-round, but with some degree of seasonality. The life span of the animal is about 10 to 40 years. Reddis generally do not hunt this animal because they don't consume the meat of the animal. According to Reddis, the footprint of monkeys resemble the hind leg of porcupine.

They identify another species called *Bandalu*. This animal resembles monkeys but the body colour is grey and the face, hands, eyes are black. This animal lives in deciduous forests. The animals tail is double the length of a monkey's tail. The food habits and reproduction system is similar to the monkeys. According to *Reddis,* the mating season of the primates generally is during the onset of the monsoon. Footprint of Bandalu is same as that of monkey but slight bigger. Konda Reddis do not eat the meat of *Bandalu*.

Felines in the Forest

Rekka pilli: This animal looks like a cat but has white wings. Wings are useless and the animals move slowly on the trees. The height of an adult animal is one feet and weights approximately 5 kgs. It is a forest-dwelling species and lives in all types of forests. It is nocturnal, terrestrial, sedentary and solitary but occasionally lives in pairs and is apparently territorial. It is usually found on big tress and also in sacred groves. They remain inactive during the day in their underground burrows or in the hollows of fallen trees in swampy areas. During the night they follow their own pathways through the forest in search of fruit and seeds. According to the Reddis, it is a very cunning natured animal that lives on sacred trees. This is because it knows that people cannot climb on the sacred trees for hunting. The animal can also climb big trees up to the tip of the branch and can also jump from any height and uses the wing to glide. Reddis say that it is an endangered species, which they spot very rarely in the forest and also they do not know much about its food habits, mating habits and other activities. Some people said that they have not seen this animal for years. Its meat is edible.

Bootha pilli: (Felis chaus) these forest-dwelling species can be found in all types of forest. It is black and white with a long tail and looks like a cat. The hiding place of the animal is bushes preferably in the vicinity of water and in cultivated patches. These animals are nocturnal and are most active at night. Size and weight vary by food. Total adult height is one feet and adult weight is 3-6 kgs. They categorize the male as *Pothu* and female *Adadi*. Footprint of the animal is similar to dog but fingers are very close and short. The animal's excreta resembles that of cat. An important behaviour of the animal is that at the time of catching the prey it walks very slowly without making any noise. If it spots any human being it just runs away.

It feeds on a great variety of aquatic animals, insects, files, ants etc. These animals eat fish (*Chapalu*), snails (*Nattagulallu*), crabs, (*Pithalu*), all types of insects (*Purugulu*), flies (*Egalu*) and salty soil (*Soudu*). According to information, the life span of the animal is not as much as expected because of food habits.

It takes a lot of non-vegetarian food and accumulates fat and faces lot of problems with the fat content in the body.

Sexual maturity occurs at the end of the first year of life. Mating season of the animal is in August and babies are born in the winter season. It gives birth to 2-4 babies once in a year. The animal can reproduce babies till ten years.

Thom pilli: This animal looks like a cat and the eyes of the animal are blue tinged with grey. It has dark dense moustaches, wings and a long tail of about one meter. The body colour of this animal is black and it is about one foot tall. The weight of an adult animal is around 3 kgs. It lives in groups consisting of adults of both sexes and their offspring. The hiding place of the animal is holes of tree (Chetla thorallu), rocks, (Rallalo) etc. When Reddis go for hunting they find these animals on rocks and while they sleep in the holes of trees. Thom pilli builds nests on the trees to live and to give birth to babies. Generally this animal is not aggressive but attacks if teased or if its babies are threatened.

These animals consume vast quantities of plant food. They feed on flowers, tubers and fruits like, *Pala pandlu, Busi pandlu, Mangoes (Mamidi pandlu), Nakkori Juvvi pandlu, Glassa pandlu, Parmi, Jerrimudu, Ravi pandlu, etc.* They eat the flowers along with the fruits of the trees like Neem flowers *(Vippa puvvu), Chilakadidiri, Nallgidi Burugu.* They also eat *Soudu,* insects like *Tummeda,* scorpio *(tellu),* millipedi *(Jerri), Cheda purugulu, Ladde purugu, Pasara purugu,* butterfly *(Gaga),* and tender leaves, salty clay (*Soudu*), animals like Rats *(eluka),* rotten flesh of the dead animals, flies etc. The special character of these animals is that they drink *Zeelugu* toddy in the night. They eat whatever food is available completely and then look for something else. They do not harm the crops.

Young animals acquire adulthood in about one year. Mating season of the animal is in summer and the babies are born in June to August. They nest on trees for babies. They give birth to an average of two babies once a year and continue reproduction up to 6-8 years. Hunters wait near fruit trees and other plots for hunting this animal. Reddis eat the meat of this animal and it said to be very tasty.

Bentuku pilli: These are found in a great variety of habitats from forest and bush land to swamps. *Reddis* say that this animal looks like *Mungisi pilli* and an adult animal weighs around 2-3kgs. Their living and hiding place is in holes, on the tree trunks (*Thorallu*) and in the burrows (*Randralu*). Reddis hunt this animal for consumption. The animal is very scared of snakes and runs away when it encounters snakes.

Kurude pilli: It is a forest-dwelling species lives in all types of forests. This animal resembles an young wild pig and has three white stripes on the body from head to tail like a squirrel. The male (*Pothu*) has small tusks protruding out of the corners of the mouth and the female (*Adadi)* does not have a tusk. The skin of the animal is thick and strong. The weight of an adult animal is around 5 kgs. Reddis say that the hiding place of the animal is in dense bushes *(tupallu*), rocks, *Chitti adavi* or *Kondaothu*. The animal shies away from the humans. These animals are nocturnal, terrestrial, sedentary and solitary. They occasionally live in pairs and are apparently territorial. These animals use tusks for self-protection and attack if they sense any danger. A peculiar behaviour of the animal is that it can jump from any height for self-protection.

The Food habit of the animal is purely herbivorous. It feeds on a great variety of tree fruits, leaves and seeds. According to Reddis this animal eats fruits like *parmi, Nakari, Guluju, Vokapandlu* and grass like *Garki gaddi, Tungagaddi, Sikum* (tender leaves), pulses in the field, *Soudu* etc.

Sexual maturity occurs at the end of the first year of life. Mating season of the animal is in summer and the baby is born during the month of July to August. It gives birth to two babies in a year and can reproduce up to 5-6 years. It is mainly hunted at night with the help of torchlight and the meat of the animal is very tasty.

Pandu pilli: This forest-dwelling species can be found in all types of forests. It is red with a long tail and is similar to *Thompilli*. It is a herbivorous animal and food of the animal is fruits like *Ippa pandlu, Adda pikkelu, soudu* etc. Its hiding place is rocks (*PeddaRadlu*) and the cat behave similar to *Thompilli*.

Putta pilli: These forest-dwelling species preferably live in the vicinity of water, and in cultivated patches. The animal has long tail and white spots on the body like leopard. These are nocturnal, terrestrial, sedentary and solitary. They occasionally live in pairs and are apparently territorial. They remain inactive during the day in their underground burrows or in the hollow hole of fallen trees in swampy areas. During nights, they follow their own pathways through the forest in search of food. The hiding place of the animal is *Puttalu, Thoraalu* and the bushes at canal side. They eat the stems, roots, tubers, grass and aquatic animals. The food of the animal is grass, paddy, snails *(nattagulalu)*, fish *(Chapalu)*, crabs (*Pithalu)* etc. there is variation in their diet in different places and at different times depending on what plants are available and the stage of the growing cycle. *Reddis* often spot this animal in the fields and forest. Hunting along canals is probably the most common method for Reddis. They eat the meat of this animal very often and catch it by trapping (*uchu*).

Bouru pilli: This animal is a forest-dwelling species that lives in all types of forests. According to Reddis this animal resembles the domestic cat, except that these are bigger in size. The body is black in colour with gray patches all over the body. An adult animal weighs around 5 kgs. Reddis say that the hiding place of the animal is thick grasses, rocks, holes in the tree trunk (thorralu), banks of canals and streams (Gattlu), and on the trees (chettlu) etc. It remains inactive during the day in their underground burrows or in the hollow holes of fallen trees in swampy areas. It hunts birds on the trees. Sometimes stray into villages and snatch away fowls for food.

This animal feeds on a great variety of vertebrates like, birds, poultry, rabbit and Soudu (salt lick) etc. It may prey upon domestic animals like poultry. Reddis believe that sexual maturity occurs at the end of the first year of life. Mating season of the animal is in summer and the babies are born during the month of August and the animal can give birth to babies for around ten years. It is mainly hunted at night with the help torchlight. Trails and traps are also used. It is mostly hunted for domestic consumption

Rodents in the Forest

Rasi eluka (Funambulus pennanti): This animal is a nocturnal animal. The tail of the animal is longer than the body and it resembles a rat. The special character of the animal is that it has a 5 or 6 colour stripe on the body (i.e. green, yellow, blue, red, black white). An average adult animal weighs around 3kgs. The animal is found on big trees like *Burugu* etc. An identifiable character of this animal is that it builds a nest to live and reproduce in it and is always in a pair.

The average weight of the animal is 2-3 kgs. It builds a big nest on the tree and lives in groups. It drinks water from the *Burugu* flower, which is very sweet in taste. It is a purely herbivorous animal that feeds on a great variety of tree fruits and seeds like *Kalulu, Addapapu, Usirkayalu, Jedipandlu, Mamidi pandlu, Busi pandlu* etc. In summer after eating food it sucks nectar or water from *burugu* flower.

Mating season of the animal is April to May and the young animals are given birth during July to August in the nest on a tree. They attain sexual maturity at the age of 2 to 3 months. Reddis are very fond of its meat, which is very tasty.

Sunchieluka (Rattus rattus narbadae): This is a small size house rat found in human dwellings and their surroundings. It is often available in kitchen and storerooms. The colour of the rat is black. This animal is nocturnal and comes into the field and houses for food. The rat weighs about 50 grams. It is considered as a harmful rodent as it feeds on and damages all food varieties and crop available in kitchen and storerooms. This rat primarily feeds on crop and all cooked food varieties. Reddis do not know much about the mating season of this animal. But the babies of the animal are seen more often in summer season. *Reddis* do not kill or eat this animal.

Gadde eluka: This is a big size rodent, commonly found in nearby granaries (*Gadde*). The colour of the animal is black. This animal is nocturnal and comes into the agricultural fields for food. The storing place is locally named as *Gadde* where Reddis store paddy and other crops and the rat's name is derived from this. This animal is considered harmful as it feeds on and damages all food grains like paddy and seeds, which are stored in Gadde. The hiding place of the animal is holes, which are found in and outside the village and mostly in the fields. General character of the animal is that it feeds on stored food grains, seeds, fruits and vegetables in the daytime and hunts the prey in the nights. Reddis do not eat this animal.

Yelita eluka (Milardia meltada): This is a small size rat that resembles *Suri eluka.* But the end of its tail is thicker with hair. This animal is black in colour, commonly found in burrows in the soft soil on the corner of the fields and banks of canals. This animal is nocturnal and comes into the agricultural fields for the food. This animal weighs around 100grams. The general character of this animal is that it lives in groups of 5-15. It feeds on very small size fruits, insects, seeds, corns like Jowar corns (*Jonnaginjelu*), cottonseeds (*Pattiginjelu),* groundnuts *(Veruchanagalu),* etc. *Reddis* do not eat this animal.

Mongoose or Herpestes edwardsi (Mungisi pilli or Chimala mingu): These are found in a great variety of habitats from forest and bush land to swamps. The fur of this animal is light brown and black in colour. It has a long bushy tail and the adult weighs around 1-2 kgs. Reddis say that the living and hiding place of the animal is in the big bushes (*Donka)* and near the anthills (*Puttalu).* It eats snakes, rats and birds eggs, flies (*Egalu),* insects, birds *(Sachi poina pittalu)* and carcasses (*Kalebaram*) of dead animals. Mating season of the animal is not known to the Reddis. Special character of the animal is that they fight with snakes. Reddis do not eat this animal flesh.

Squirrel or Funambulus palmarum (Udutha eluka): This animal resembles a rat but it has a bushy tail with hair and has three white stripes on the body. The colour of the animal is grey. The animal is commonly found in the forest and villages. An identifiable character of this animal is that it builds a nest on the tree to live and reproduce in it and is always in pairs. The animal weighs around 100grams. The animal feeds on fruits like mangoes *(Mamidi*), coconut (*Kobari)* coconut water, *Dosa pandlu, Bobari,* tamarind *(Chintakaya)* and some other wild fruits. Hiding place of the animal is bushes or on the trees. *Reddis* do not know much about the mating season of this animal, but the babies of the animal are seen in summer. Reddis do not eat this animal meat.

Canines in the Forest

Fox or Canis lupus (Nakka): This is a wild animal similar to dog, with pointed face and ears, a wide fury tail and reddish brown fur. The tail is locally called Gujjuga. Reddis categorize fox into three types on the basis of their size and age. *Nakka* is a bigger one and old aged, *Gunta Nakka* middle sized and middle aged, and *Konka Nakka,* is the smallest young one. A small animal weighs 35-40 kgs and the bigger ones weigh up to 70 kgs. Among the three varieties, *Konka Nakka* is very dangerous. If it confronts a human, it directly attacks the lower part of the legs. The lower part (anklet) of the leg is locally called as *Konka.* They may travel kilometers every night or they may remain for two or three days near a recent kill. The hiding place of the animal is caves, rocks in dense forest, canal banks, and field bushes. *Reddis* name male fox as *Pothu* and female as *Adadi.* The footprint of animal resembles that of a dog but the difference is that the fox footprint is small in size. When Reddis go for hunting, they recognize the animal by the bad smell of its wind.

The fox eats a great variety of vertebrates like rodents, rabbits, deer, large birds etc. In inhabited areas it may prey upon domestic animals like pigs, dogs, poultry (*Koddlu),* tortoise (*Thabelu),* fish (*Chepalu), Soudu* and even full-grown cattle. It also eat fruits like *Dosa pandlu, Pucha kayalu, Jonna pothulu,* ground nuts (*Veruchanagelu*) etc.

Young animals reach sexual maturity in one year and they mate in the summer season. The gestation period of the female animal is about 7 months. A female fox gives birth to 4-5 babies in a litter. The capability of reproduction is up to ten years then it becomes old.

Manupuli: According to Reddis, this animal is rarely visible and found in the holes of the tree trunk. The animal is considered harmful and attacks cattle. Konda Reddis say that this animal lifts its prey by the tail and sucks the blood.

Konda Reddis identify wild animals in the forest by listening to their sounds, foot prints, blood spots when they are injured, ditches and burrows dug by the animal in the ground for food and shelter, rubbing spot where the animals rub their body to get relief from itching, etc. *Reddis* say that some of the animals, birds, insects and snakes eat *Soudu* (salty soil) for digestion after eating their food.

Reptiles in the Forest

Ground lizard or Sitana ponticeriana or Varanous monitor (Udumu): This reptile is big in size and has long body, four short legs, long tail and thick skin. It is a forest-dwelling species of both hot and moist types. According to *Reddis*, this lives mostly in holes *(Randralu)*, in the tree trunks *(Chettu thorralu)*, anthills (*Puttalu)* and in stone caves. The average weight of an adult animal is approximately up to 5 kgs. This animal is generally seen in summer season and it has a distinguished forked tongue. Reddis differentiate males as having two testicles and call it as *Pothu* where as the female is called *Adadi*. Reddis categorize this animal into three types depending on its size. They are *Chemiti udumu, black* in colour, small in size, with grey dots on the body and weights around half kg. Second one is *Langa udumu,* medium in size; dark black colour and average weight around 1-2 kgs. The third one is *Gali udumu,* light brown in colour and weights 2-3 kgs. The general behaviour of the animal is that it continues to sleep in the same spot.

These are nocturnal, terrestrial, and solitary animals that occasionally live in pairs and are apparently territorial. They remain inactive during the day in their underground burrows or in the hollows of fallen trees in swampy areas. During the night they follow their own pathways through the forest in search of fruit and seeds. They feed on a great variety of tree fruits and seeds. They seem to be quite unselective and the main food in their diet may vary from one place to the next. When fruit is scarce they have their diet of shoots and leaves, sometimes helping themselves to various crops. They feed on insects like *Tummedapurugu, Chedapurugulu, Errapurugulu, crop insects, butterflies, tender grass, Soudu and Nakoripapu* etc. The seasonal scarcity of fruit from November to January is a major limiting factor for animals to eat crops in the fields.

Reddis believe that their sexual maturity occurs at the end of the first year of life. Mating season is with the onset of monsoon. A females lays 50-100 eggs in holes, which are hatched during July-August. The eggs are the size of grapes. A female animal can lay eggs up to 15 years.

It frequently features in the list of subsistence hunting game animals. The hunters post themselves around areas where fruit fall is heavy or where there are crops these animals like to raid. They are also hunted with dogs, which drive them into burrows or hiding areas. *Reddis* are very much fond of its meat and eggs. Meat is of good quality and also tasty. They also say that if the curry is cooked in the morning it gives a very good taste in the evening. People say that by eating this flesh, body and waist pains are cured.

Monitor lizard or Varanus bengalensis (Valugu): This reptile has long body, four short legs and thick skin with long tail. The head of the animal is very small and the body is large covered with scales. The scales of the animal are very strong. The tongue of this animal is very long and is coiled. According to Reddis the animal resembles Udumu and dinosaur. The animal is nocturnal and lives in very moist places. The hiding place of the animal is on rocks and also found in the burrows. Since its mouth is pointed and sharp, it can easily make burrows in the ground. It is expert in making holes. An adult animal weighs about 10-15 kgs and length of the animal is about one meter long. The animal varies by size, weight, sex and age. The footprint of the animal resembles that of an Udumu, but the difference is that its nails are bigger in size.

Reddis categorize this animal into three types according to the size. *Konda valugu* is grey, very large in size and about four feet long. Second one is *Maddi Valugu* or *Nalla valugu, is* black, medium in size and about three feet long, weights about six kgs with whitish patches (chippalu) on the body. Third one is *Erra valugu* or Kodivalugu is red, small in size; it is two feet long with whitish patches on the body. The animal comes out in the night for food in the fields. Hides in the holes and on the rocks of other animals. The footprints are similar to *Udumu* but difference is *Valugu* is that its nails are bigger in size.

The animal feeds on ants, aquatic and terrestrial insects, like Chemalu *(ants)*, Chada purugulu *(anthills insects)* Nathagullalu *(snails)*, Thummeda purugulu *(wasps)*, Gumediginja *(seeds of ash gourd)*, Soudu, *frogs* (Kappalu), *crabs* (Pithal), fruits like bodda pandlu, Nerodi pandlu (Zizupus sp.). The basic adult diet varies from place to place and includes fish in summer. Some times this animal strays into the fields for food. If the *Konda valugu* is very hungry, it also eats two varieties of infants i.e. *Kodi valugu* and *Maddi valugu*.

Mating is during the summer season and birth of younger ones coincides with the onset of the season. A month or two later, the females lay a clutch of 14 to 76 eggs weighing 9-14 gm each in burrows excavated in communal nesting sites. At the end of a three-month incubation period, the newly hatched animals come out of the nest.

They are most frequently hunted for food. The methods used for hunting is with dogs that are locally trained to capture by shaking the lizards out of their trees. *Reddis* eat the meat of this animal, which tastes like poultry. The eggs are also gathered and eaten. According to the Reddis, eating the meat of this animal has made them strong and powerful.

Snakes (Pamulu)

Snakes and lizards are both highly diverse and particularly abundant and variable. In addition to their myriad functions within their respective ecosystems, the most valuable species are also fairly important and are discussed below.

Reddis identify different species *(Jati)*, of snakes and they call snake as *Pamu* in their local dialect. *Reddis* categorize snakes into poisonous (*Vishamugala)* and non-poisonous (*Vishamulenidi*). *Konda Reddis* classify the snakes by various criteria such as the size, sex, hiding place, body colour, location food habit, predating habit and special characters of the snakes. They can easily identify a number of snakes in the forest and their surroundings. They give clear description of each of these snake varieties. Some of these descriptions are given below.

Poisonous Snakes (*Vishamugala pamu*)

Nagu pamu (Cobra or Naja naja): It is long, black in colour with a white stripe on the body. Its distinct feature is that it has a hood *(padaka)*. It is a poisonous snake. Snakes attack if threatened. It can attack on any part of the body in any direction. If it bites, froth comes out of the victim's mouth and if not treated on time it can be fatal. This snake feeds on insects (*purugulu*), frogs (*kappalu*), wild rats in the forest, salty clay (*Saudu*) and small size reptiles and aquatic animals. Snakes live in thick bushes in the forest and fields, anthill, in the weeds (*Donkalu*), tree trunks (*chettla thorralu*) and sometimes on the trees. Reddis say that these snakes are rarely seen as they shy away from the people. Reddis do not know the exact reason of the mating of the snake, but they find the young ones in the forest and fields during the onset of the monsoon.

Nalla Thasu (king cobra): It is long, black in colour and resembles cobra. It is more poisonous than cobra and has similar habitats and foods.

Poda: It has a flat long structure, a big head and a blunt tail. This snake feeds on insects, frogs, *Soudu* and other small size animals in forest. It is a poisonous snake. Snakes attack if threatened. It can attack on any part of the body in any direction. If it bites, froth comes out of victim's mouth and if not treated on time it can be fatal. Mating season of the snake is in summer and young ones are seen during onset of the monsoon.

Guddi Poda (blind snake) Typhlops beddomei: It is long, white in colour with a black stripe on the body. It is a nocturnal snake. *Reddis* say that this snake snores in the night. It is as poisonous as cobra. This snake also flies in the air. According to Reddis, this snake mostly comes out in the rainy season.

Raghutha Poda (viper) Vipera russeli: It is long, whitish in colour with a slightly flat body. If it bites, blood comes out of victim's mouth and if not treated in time it can be fatal. This snake can fly in the air.

Konda Poda: It is long, whitish in colour with a black stripe on the body. It is as poisonous as cobra and has similar habitats and food. If it bites, froth comes out of victim's mouth and if not treated in time it can be fatal. This snake can fly in the air.

According to *Reddis*, this snake mostly comes out in the rainy season.

Godum Thasu: It is long, whitish in colour and resembles cobra. It is a poisonous snake. If it bites, froth comes out of victim's mouth and if not treated on time it can be fatal. According to Reddis, this snake is mostly seen in summer.

Akudolla (green snake) Dryophis pulrerulentis: It is slim, long, green in colour with small size head. It is a poisonous snake. If it bites, froth comes out of the victim's mouth and if not treated on time it can be fatal.

Giri nagu: It is long and pure black in colour. It lays eggs in the nest on the trees.

Thadelu (Krait) Bungarus bungarus: It is long, back in colour with white stripe on the body. It is more poisonous than cobra. According to *Reddis*, this snake is seen through out the year.

Gotte Poda: It is small sized, whitish in colour with white spots on the body.

Chakka pamu: It is long, whitish in colour with black spots on the body. Its distinct feature is that it has a hood (Padaka).

Dhul Nagu: It is small sized, whitish in colour with white stripe on the body. Its distinct feature is that it flies in the air and hits the head.

Naga Jerri: It is long, black in colour and resembles *Nalla Jerri Pothu.*

Pinjerga: It is a long snake, green in colour. Its distinct feature is that it has a hood (*Padaka*).

Kodathasu: It is long and black in colour.

In all, Reddis identified 16 poisonous snakes along with their hideouts and food habits like consuming rats, frogs, small insects, lizards and other animals in the forest. Living and hiding place of the snakes are in the anthill, bushes and in the leaves of the trees.

Non-poisonous Snakes (*Vishamuleni pamu*)

Erra Jerri Pothu: It is a very long snake about 3 meters and yellow in colour.

Nalla Jerri Pothu: It is a long snake, black in colour and resembles *Errajerripothu.*

Vapu dolla: It is a small sized, snake, whitish in colour with a big head. It has a prominent black nostril and hind limbs. It feeds on green leaves, grasses and small insects. According to Reddis, it is mostly found on *oddisse* tree and in the green grasses. If the snake crawls on the human body, the whole body starts itching.

Vana pamu: It is long, whitish in colour with black stripe on the body. It usually feeds on frogs, water insects, fishes, *Soudu*, etc. Its living and hiding place is water bodies and bushes. *Reddis* say that this snake is a good swimmer and spends most of its time in water. This snake is mostly seen in the rainy season.

Konda chelava (python) Python molurus: It is a very long and big sized snake. The colour of the snake is whitish. It can swallow a human being. The living and hiding place of the snake is very big trees, on the rocks and dense forest. This snake is rarely seen by the *Reddis*.

Burada Pamu: It is long and black in colour. It can be found in the bushes and corners of the water bodies.

Chintha Neppu: It is small sized, whitish in colour with black spots on the head. It can fly in the air.

Pachirga pamu: It is a long, pure green in colour with big nostrils. It is found near water bodies and in the bushes.

Palakarra Mondadi: It is small sized snake, white in colour with a blunt tail. It is very scary. This snake feeds on milk of women. It is mostly found in near by bushes and on the roof of house. *Konda Reddis* believe that it comes in the night to drink milk from lactating mothers. Sometimes it sucks milk from the lactating cattle and goats as well.

Jergoddu: It is long and light red in colour.

Immedula: It is long and light red colour and mostly found in nearby water bodies.

Reddis identified 11 non-poisonous snakes along with their hideouts, foods habits like insects, ants, rats, frogs, lizards, *Soudu* and small animals in the forest. They live in anthill (*Puttalu*), bushes (*Tupalu*) and also on the trees.

Knowledge about Diseases of wild Animals

Generally, Reddis do not know much about the wild animals' diseases. But they say that the common diseases that afflict the domestic animals can be found also among the wild animals. They come to know about the disease of an animal when they hunt it or through observation from a distance. The common diseases of the wild animals which are known by the Reddis are *Kadla gadlu* (foot disease), *Noti gadlu* (mouth diseases), *Morsha* (epileptic fits), *Rasapundu* (vertebral diseases), etc. A disease that is specifically found in rabbits (*Chevulapilli*) is epileptic fits (*Morsha*), because of which, pregnant women are advised not to eat the meat of rabbit suffering from *Morsha*. According to Reddis *Rasapundu* disease is found among the *Kanasu* at the neck region. This animal gets other diseases like *Kadla gadlu* on its limbs and *Noti Gadlu* at the corner of the mouth. Wild boar (*Adavi pandi*) is afflicted by diseases like *Kadla gadlu* on its limbs. *Reddis* say that if any person is suffering from any diseases they should not eat the meat of *Kanasu, Pandi* and *Chevulapilli*. If anybody eats the meat of diseased animals, the person might die. They also say that people do not eat the meat of the animal, which is attacked by a tiger. If they eat it they could be afflicted by the disease called *puligaddalu* that attacks neck.

Birds

Reddis identify different species (*Jati*) of birds. Reddis call a bird as *Pakshi* in their local dialect. They can easily identify more than 70 species of birds both flying (*Yegare*) and flightless (*Yegarani*). They are aware of their types, places of shelter, general pattern of behaviour, movement and their availability. The following list of birds is only representative and not an exhaustive one. These are birds identified by Reddis in the vicinity of their village.

Numerically speaking, birds head the list of game animals. They are found in a wide variety of habitats with the greatest diversity in the region. The most widely distributed species are discussed in the chapter. They are mainly hunted by subsistence hunters, who lure them by imitating their calls. The quality of their meat and their lure as game birds make them very attractive to the hunters. The fact that most species are quite small dampens the attraction for subsistence hunters.

The range of birds considered as game species varies regionally in accordance with the faunal composition, traditional patterns of use and types of users. Traditional game birds described below may be important regionally, particularly for Reddi hunters. Some of the birds, which Reddis mostly hunt, are mentioned below. They, eat the eggs of some of these birds. They hunt birds in the nights. Arrows, nets, traps and other devices are used for bird hunting. Although the Reddis' knowledge about birds is quite extensive, they talk about the following favourite species of theirs. Some of these descriptions are exemplary evidences of their knowledge about the birds, which they usually hunt.

Rushi pitta: These birds are nocturnal and resemble bats (*Gabbilam*). The wings of the birds have claws with which they hang upside down from the branches of trees. They are dark brown in colour. These birds feed on fruits, pulses and other crops. They are fond of the fruits of toddy palm (*Tati pandlu*). They also consume toddy (*Tati kallu*) from the pots meant for toddy collection and sometimes the toddy gets spoiled because of the birds' defecation. According to Reddis, a distinct feature of these birds is that they feed continuously for one month without defecating and during the next month they do not eat anything but continuously defecate. These birds glide like an aeroplane. These birds are viviparous and directly give birth to the young ones. It is said that if the excreta of these birds falls on the toddy palm, the leaves fall down. Similarly if the excreta of these birds if by chance falls on the human head and is not cleaned immediately, the hair will be lost. These birds flock around the sources of water in summers. Konda Reddis do not eat this bird. They are herbivorous.

Titva pitta: The body of this bird is light red on the upper side and white on the lower side with long legs. The nose and legs are yellow in colour. They build nests on the rocks and in the bushes. These birds feed on insects like grasshoppers (*Medatahlu*), *Gaga*, *Chedapurugulu*, crop seeds, fruits and pulses. These birds have a crest on the head. The beak of the bird is small in size and yellow in colour. Sometimes if they feel hungry, they eat their own eggs. A distinct character of these birds is that on seeing wild animals in the forest they make a sound, which alerts the Konda Reddis while they hunt. These birds are more useful especially during the night to identify the animal. Mating is around the onset of the rainy season. These birds lay 5 eggs on the ground or on the rocks i.e. (*Dibba*) a two-sided slope and at the tip of the rock. The eggs have black spots.

Chambaru burka pitta: This bird is black in colour with red tail. The adult weighs around ½ kg. These types of birds are mostly found in *Chitti adavi* on the top of the trees and build nest on the trees. This does not allow any body to the nest. These birds feed on domestic and wild fruits, grains, fruit seeds, insects, flies, etc. It is a very dangerous bird. The hiding place of the bird is thick tendril *(tutlu)* in the forest. The bird is most intensively hunted species in the forest. They are seen in pairs or in small groups on the top of the trees but may occasionally be seen in the underbrush. They are most active in the morning and afternoon. Relatively shy when detected they do not fly far but try to hide in the high, leafy treetops.

This bird lays four pure white coloured eggs and hatches four eggs, which are incubated for 15 days by the female. When they have young ones, both male and female feed them. The young ones are ready to give birth at the age of four months. The female bird can give birth to the young ones till it attains an age of 15 years. They are stalked in the underbrush along the forest trails, hunted from along rivers and located by their calls, or hunters may lurk near their drinking, eating or resting places. Reddis eat these birds meat and they believe that their meat has medicinal properties. They said that the meat of this bird could cure diseases like *Vatam* (body swelling) and back pain.

Endritha pitta: They live alone or in groups of four to twelve, rarely in large flocks. It is pure black and the tail resembles the fish tail, very small and the bird resembles *binge raju* bird, but it does not have head, beak and tail. The weight of the bird is around fifty grams. The hiding place of the birds is *Chetla Thoralu,* on the tree and also in the bushes. It is nocturnal and diurnal and hunts for food during day and night. It hides away the chicks from the other wild animals and it always fights with other birds.

It feeds primarily on crops, pest in the fields and insects. Some times it sits on the domestic animal and searches for food. It builds a simple nest in the trees and generally lays a clutch of five eggs; eggs are in white colour, which are incubated by the female for 24-28 days. Generally it is very active in early morning around three thirty and conveys the message to the people to wake up.

Pachi endritha pitta: These birds are small in size and weigh around 6 grams. These birds are green in colour with black nose and legs. It is very difficult to identify male and female because they look similar. A distinct feature of these birds is that it makes noise like *endritha pitta* and therefore it is named as *pachi endritha*. These birds feed on insects, flies, fruits and crops. These birds lay 2 white coloured eggs and hatch two.

Batti pitta: According to *Reddis*, this bird is golden and grey coloured. It is white coloured at the tip of its tail and wings. The bird's nose and legs are white in colour. The bird resembles *Mainagora* bird. It is occasionally found in the forest. It is seen in all seasons but is more often seen in summer. The weight of the bird is about 100 grams. They live in burrows, in the trees trunks (*Chetla thorallu*) and on trees in the nest. *Reddis* mostly see this bird on the borassus tree (*Tati chettu*). It spends less time on the trees, frequently feeding in the underbrush and may walk long distances for drinking water. The bird is shy and cautious and seeks protection in the tree holes when disturbed. It is omnivorous and primarily feeds on fruits, seeds, pulses, food grains, buds, tender leaves and insects like *Cheda purugulu, Midatha, Ladde, Pasara* etc. These birds move around in search of food in groups.

They lay eggs in the holes of the trees and also in the nest. They lay around four eggs each time. Eggs are green in colour with black spots. These birds are early risers and they are very caring and protective of their hatchlings. Generally, they lay a clutch of four eggs and hatch four eggs, which are incubated by the female for 24-28 days. The young ones are ready to give birth at the age of four months. When they have young ones, all the birds will fly in a flock for getting food for themselves and the young ones. Generally the birds are active in the morning and noon. *Reddis* do not eat this bird. They hunt to keep them as their pets. A distinct feature of this bird is that on training, it repeats the words spoken to it.

Bingiraju: The bird has a pure black and light red colour coat. The tail resembles a fish tail, where two long feathers fork out. It has big sized eyebrows. It has a crown on the head, called *Koppu* in local dialect. Two long feathers protrude to form the wings of these birds, called *Gujiga*. The adult bird weighs around 100 grams. These birds are seen in all seasons and more often in the summer season. They spend less time on the trees than other birds. These birds move around in search of food in groups. A distinct feature of these birds is that they can repeat the words, if given proper training. These birds are omnivorous and primarily feed on fruits (*Tatipandlu, Marri)*, seeds, pulses, food grains, buds, tender leaves and insects like butterfly, flies, mosquitoes etc.

These birds build nests on trees and bushes and lay eggs, which are grey coloured with black spots. They lay around 4 eggs, which are incubated by the female for 24-28 days. The young ones are ready to give birth at the age of four months. The female bird can give birth to young ones till it attains an age of 20 years. When they have young ones, all the birds fly in a flock for getting food for themselves and the young ones. Generally, the birds are active in the morning and noon. It is said that these birds help *Reddis* in hunting, since they know where the wild animals hide. Reddis do not eat this bird. They catch to keep them as their pets.

Mainagora: These birds are pure black in colour with white wings, and nose and legs are yellow in colour. These birds live in trunks of the trees (*Chetla thoraalu*) and in the bushes. These birds are seen in summer and winter seasons. The adult bird is about 100 grams. They are purely herbivorous. They feed primarily on fruits and seeds like *Neradi pandlu, Alle, Juvvi, Pardonda, Korkudu, Busi, Marri, Rege, Glasspandlu, Raavi, Bankanokari, Parim, Pallori, Chilakadidiri, Vyapa, Giriga, pedda pala,* and other grains. These birds go around in flocks for food.

They lay four eggs in the trunk of the trees in summer, which are grey in colour with black spots. It is said that only 2 eggs are hatched and other two are left to rot. These birds hatch the eggs within 15 days.

These birds can also speak and they also recognize the voice of a person and can imitate the voice. A distinct feature about these birds is that they dip in water and take bath everyday and also sleep in the same place every day. If any body tames this bird, they have to be given bath everyday with turmeric to ward off evil eye (*Disthi*). And also the whole family has to take bath everyday. If they do not take bath every day the bird flies from his place. They are found in forest. Outside the pairing season, they might band together in flocks of ten or fifteen. They often sleep at night in permanent nests in trees along the riverbank. They are silent and timid. They are mostly available in summer and winter season. *Reddis* do not eat this bird. They catch to keep them as their pets.

Gunjipitta or Gudlaguba: This bird is grey in colour with black feathers. These birds have large eyes and head, sharp and small nose, claws, long and sharp nails. The weight of the adult bird is about 200 grams. It is a nocturnal bird. These birds are seen throughout the year and they live in the trunk (*chettla thorallu*) of the trees. They are rarely seen in-groups and they often sleep at daytime. These birds have different styles of flight. When the birds fly slow and low, they flap their wings and when they reach high in the sky they glide and do not flap wings. They are omnivorous and their diet consists of seeds, roots, aquatic animals, insects *(Purugulu)*, flies *(Egalu)*, rat, dead animals and other small animals. Sometimes in case of extreme hunger the bird kills and eats its own babies. It lays two eggs, which are pure white in colour, and hatches both of them. These are incubated for 25 days by the female.

Tevari buchipitta: The body colour of the bird is grey, wings are green and nose and legs are white in colour. The bird resembles *Pala Pitta*. A distinct feature of this bird is that they make different types of sounds. It saves chicks from other birds and also flies in different styles. They fly rarely in a flock. The bird is available throughout the year. The hiding and living place of the bird is *chetla kommalu,* on the trees and bushes, holes of trees. The weight of the bird is around 300 grams. Reddis do not eat this bird.

The bird is omnivorous and its diet consists of seeds, fruits, roots, aquatic plants, crops, insects, fish and other invertebrates. Nesting, in hollow trees and mating mostly at the onset of the rainy season. It lays 4 eggs; eggs are in white colour, which are incubated for 15 days by the female.

Polugu Pitta: This bird is big in size and Reddis group this bird into 2 varieties based on its colour. While *Kaki polugu* is grey in colour with green wings, *Pacha polugu* is light yellow in colour. In these, there are two types of birds; one is *Kaki polugu,* which is ash colour with green wings and the second one *Pacha poplugu,* which is light yellow in colour. The weight of the bird is around one kg. They inhabit around rivers and sometimes in very tiny bodies of water in forest and on the trees. The hiding place of these birds is big sized trees near water bodies. Special feature of the bird is that they are very noisy and they always make noise near the water bodies. Reddis are very much fond of this bird's meat.

They are omnivorous, their diet consist of fruits, grains, insects, fish, other aquatic animals, etc. The two species have rather similar diets and feed on easily available stuff.

They nest during the rainy season from June to August and mate mostly at the onset of the rainy season. The female bird lays two eggs, which are white in colour. The chicks reach adult and weight in six to seven months and can reproduce by the age of one year.

Chiluka pitta (Psittacula cremari or Parrot): Reddis identify four different types of parrots based on their size and colour. They are *Chiluka, Rambachiluka, Guddichilika, Rallechilaka or Rachilaka. Chiluka* is dark green in colour, its nose and legs are red in colour. The living and hiding place of the bird is in the holes in the trunks of borassus trees (*tati chetlu thorralu*) and the weight of the bird is around 200 grams. The bird *Rambachilaka* is bigger than *chilaka* and the weight of the bird is around 250 grams. The body of this bird is green in colour, nose and legs are red in colour and at the neck it has a red patch. The ends of wings are red in colour. The bird *Guddi chilaka* is small in size and the weight of the bird is around 150 grams. The body of bird is light yellow in colour and nose is white. This bird is always seen alone. The bird *Rachilaka* is smaller than other birds. The weight of the bird is around 100 grams.

The above four species of birds have rather similar diets. They are herbivorous; depend on seeds, fruits, nuts and crops in the fields. They are mostly available in winter season. They nest, in hollow trees and apparently mate mostly at the onset of the rainy season

Tenepitta: These birds pure black with white spots. They feed on honeybees, aquatic grass, insects, pulses, seeds and other grains. They also eat insects and molluscs, particularly during the mating season. These birds are usually found near honeycombs on the tree. Reddis therefore can identify honeycomb.

Guvvapitta: Reddis categorize this into four types of species *Pidatha guvva,* small in size. *Bobberi guvva,* light red in colour and big in size. *Alubanthu guvva,* is pure white in colour and looks like *Bobberi guvva. Amida guvva* is red in colour and associated with sky blue colour, resembles *Bobberi guvva.*

Koppu pitta: The body of these birds is grey in colour with red spots. These birds have a hard crest on the head. These birds look like eagle but are bigger in size. These birds are carnivores and feed on small animals like rabbit (*Kundelu*), small sheep (*Gorre*), rats (*Elukalu*), other birds (*Pittalu*) and insects (*Purugulu*). These birds are very strong and have powerful claws with which they can kill their prey. These birds live in the thick forest and prefer to live in cool places.

Nemali pakshi (Peacock or Pavocristatus): The body of these birds is green in colour. The male bird is called *Pothu* and they have a long tail with multi coloured beautiful feathers. The female bird does not have long tail. These birds are seen among the bushes and on big rocks. These birds feed on termites (*Cheda purugulu*), paddy, *Gantelu,* fruits and insects.

Dabbi pitta: These birds are black in colour with red spots on the body. These birds have a hard crest on the head. The legs of these birds are red in colour. These birds live in the dried tree trunks (*Yendina chetla thoralu*).

Pokkeni kayapitta: The bird is green and its lower part is light red. It has a red patch on the head and long nose to make holes on the tree. It can easily be identified, as it is very noisy like *Pokke* bird. It lays eggs in the nests of dry trees.

Kancheri pitta orPedda kancheri: The bird is sky blue coloured with red wings. It has a very long nose, which is around 10-15 inches. The weight of the bird is around 2kgs. This bird is mostly seen in rainy season. It lives in big holes of the trees. It builds nests with the help of mud on huge trees. Male bird provides food at the time of laying eggs. The bird mostly feeds on bitter taste fruits like *Mushid, vyapapandlu, neradipandlu, raipandlu, alepandlu.*

Pacha pollugu: This is a large green bird. The weight of bird is around ½ kg. These birds feed on all types of fruits, which are available in the forest.

Pichuka or Panchiga (Sparrows or Passer domesticus): The bird is small grey-brown in colour. The male bird has black stripe on the lower side of the neck. These birds are seen in all the seasons.

Seema kancheri: This is a large black bird, with white wings and the adult weight of the bird is around 1 kg.

Bangaru Pichuka: The bird is golden in colour therefore it is named as *Bangaru*. A distinct feature of this bird is that it flies very slowly and build nest with the help of *Nukala gaddi* and borassus plam (*Tati chetlu*) leaves. It also prepares cradle for the chicks.

Lakullondadu: The body of this is green in colour and the nose and legs are red in colour. These birds are mostly seen in summer. The special feature of this bird is that it lives near the corner of canals on the trees and also flies very slowly.

Vankaya pandu pitta: The body of this bird is purple like brinjal, and therefore it is named as *Vankaya*. They live on the trees. The special feature of this bird is that it builds nests in winter and in summer, and also flies very slowly. Konda Reddis distinguish this birds into 2 varieties based on their colour. Other species are *Pacha pandu pitta* these birds are pure yellow in colour. They feed on all varieties of fruits, which are available in the forest.

Saakali pitta: This is a small bird, white in colour with a long tail. A special feature of this bird is that it flies very fast. These birds are mostly seen in rainy season.

Guddi sakali: The bird is light red in colour with its upper and lower part being sky blue. A distinct feature of this bird is that it swims on water surface. These birds are seen throughout the year. These birds live on trees near water bodies.

Pikkodi pitta: The bird is light red in colour. It has a crown on the head, called *Koppu*. A distinct feature of these birds is that they fly in slow motion.

Bhasha pikkodi: These birds are grey in colour. A distinct feature of these birds is that it speaks like humans therefore it is named as *Bhasha* pikkodi. These birds are purely herbivorous and feed on fruits, seeds and crops.

Purodi pitta: The bird is small, light red coloured with white scattered spots on the body. A distinct feature of these birds is that they run very fast on the ground and also sleep on the ground. Birds move around in groups. Konda Reddis say that the eggs are seen in winter season.

Gissa pitta: These birds are grey in colour. A distinct feature of these birds is that they move around in groups *(Tegalu)*, i.e. around 15-20 and they also sleep together in a queue.

Kerije pitta: The bird is large, light red coloured and similar to *purodi pitta*. The birds live in thick bushes.

Gompurodi pitta: The bird is larger than the *purodi* and *kerije pitta* birds. The bird is small, light red coloured with white scattered spots on the body. A distinct feature of these birds is they run very fast on the ground and also sleep on the ground. Birds move around in groups. These birds live in the field and forest bushes and also in the nests on the trees. These birds are omnivorous and feed on insects, flies, pulses and crop seeds.

Gabbilam (Bat or Sarcogype calvua): These birds are nocturnal and resemble *rushi pitta* bird. These birds are dark brown in colour. The head and face of the bird looks like a dog. These birds are viviparous and directly give birth to the young ones. A special character of these birds is that the wings of these birds have claws with which they hang upside down from the branches of these trees. These birds feed on mosquitoes, insects, flies, etc. These birds live in dark places like on *Tati chettu, midde (atuku)*. Young ones are seen in summer. Reddis do not eat this bird.

Kappeli or Athena brama (owl): The body of this bird is black and face of the bird is small and flat. These birds have large eyes and strong curved nails. A distinct feature of these birds is that it flies for some time, stops for a while and then flies again. This bird cannot walk on the ground. It is nocturnal. Reddis eat this bird in the night when they go for hunting. These birds live in forest (chitti adavi) on the trees in the nests. These birds are omnivorous and feed on insects, flies, pulses and crops. They lay two small sized eggs in summer.

Gumma pitta: These birds have large eyes and head, sharp and small nose, claws, and long and sharp nails. Nails are poisonous. The colour of the body is grey and feathers are black. They resemble *Gudlaguba* but are larger in size. A distinct character of these birds is that they live in cool place in the forest in the nighttime and in the daytime they live in thick bushes or on the trees. These birds are omnivorous and feed on rodents, fish, aquatic animals, small size birds and fruits. Reddis eat the meat of this bird and they eat special parts like heart, eyes, brain, etc because they are very tasty. It lays two eggs in summer.

Annor gedda: This is a multi-coloured bird that looks like an eagle. It is a nocturnal bird. The bird is seen throughout the year but generally doesn't come out from the forest. Reddis do not eat this bird. Reddis say, that if this bird comes into the village and cries in front of any house, there is a possibility of death. It is an evil bird. According to Reddis, the bird eats feathers of all the other birds' therefore it is named as *Annor gedda*. When young the bird does not have any feathers on the body and the bird starts eating feathers of other birds. These birds live in very dense forest in the bushes.

Kamledu: These birds are brown in colour and resembles crane. The nose of the bird is short. It is a nocturnal bird. These feed on insects on water surface, aquatic animals, Saudu (black salty soil). The bird is very fond of this. Reddis believe that the eggs of these birds have the properties to cure jaundice *(pacha kamerlu)* and eye disease where it is used as *katuka*. The shells of the eggs are burnt and the ash is applied to the eye for the prevention and curing of eye ailments. It lays two eggs in winter and eggs are pure white colour. The meat of this bird is very hard to eat.

Kovelupitta or kokila (Pigeon or Crocopus phoenicopterus): The bird is grey and white coloured. These birds are seen in all seasons. According to Reddis, these birds initially feed on insects, flies, and when the bird grows older it starts eating fruits like *parmi pandlu, pardonda pandlu*, etc. A distinct feature of these birds is that it lays four eggs in summer in the crows' nest, where they eat non-vegetarian food.

Kaki (crow or Corvus splendens): These birds are pure black in colour. They feed on insects, flies, fruits, crops and dead meat. Reddis do not eat this bird.

Chekku pitta: The body of these birds is red and black. While the male *Chekku pitta* bird is white, the female bird is small in size.

Pete ponnapitta: These birds are small in size and the colour of the birds are red, white, black and green.

Adavi bati pitta: These birds are big in size and the colour of the bird is black and white.

Konga (crane): These birds are big in size and the colour of the bird is pure white. The nose and tail of the bird is very long. They live on big trees in the corner of water bodies. They feed on insects; flies aquatic animals like fish, crabs, and prawns.

Kakkeri pitta: This is a multi coloured bird and the colour of the bird is white (*Telapu*), red (*Yerapu*), black (*Nalapu*), yellow (*Pasapupacha*) and green (*Akupacha*). It has a crown (*Koppu*) on the head.

Laaku mokudu pitta: These birds are big in size and the colour of the bird is red, white and black. They live in thick bushes in the forest.

Panchiga pitta: These birds are small in size and the colour of the bird is grey. Reddis see this bird very often. Reddis do not eat this bird.

Nunne batti pitta: These birds are small in size and the colour is black and white. The bird resembles *batti pitta*. The birds have oily feathers therefore it is named as *Nunne batti pitta*. A special feature of these birds is that they make speech like sounds. Reddis do not eat this bird.

Gaddi pitta: The bird is grey in colour, resembles *Pottem pitta* and the bird is around 5gm. Reddis say that it is very difficult to identify male and female because they look alike. A special feature of these birds is that they eat pests in the fields.

Jakki pitta: These birds are small in size and the colour of the bird is grey. The nose and legs of the bird are white in colour and the adult weights of bird is around 3 gm. For Reddis it is very difficult to identify male and female in these birds because they look similar.

Kamledu pitta: The bird is golden grey coloured with black nose and legs. The adult weighs around ½ kg. Reddis say the bird is very shy and it is very difficult to identify as male and female because they look similar.

Budargumma pitta: These birds are small in size and weigh around 5 gm. The birds mostly walk on the ground. Reddis say it is very difficult to identify male and female in these birds because they look similar.

Karkodi or Karukodi: The bird is red in colour with black spots on the body, similar to the poultry, *Dega kodi*. It has a crown on the head.

Koppu peckodi: The size of the bird is very small and the colour is black. It has a long crown (*Koppu*) on the head.

Ratikodi or thogari kodi: These birds are pure red in colour. It has a crown on the head and has a sharp nail on the backside of the legs called Arre in local dialect. The bird is around ½ kg.

Kanasu kodi: These birds are small in size and weigh around 300 grams. These birds are brown in colour with red coloured legs. They live on the trees in hill *(dippa)* places.

Bolli kodi: These birds are black in colour with white coloured legs. The face of the bird is similar to a donkeys face. These birds are usually found near corners of the canals in the bushes. They feed on insects, flies, fruits, crops, etc.

Thondam kodi: These birds are very big in size with very long nose.

Turkey kodi: These birds are big in size with long throat and short nose. The colour of the bird is grey. The adult weighs around 6 kgs and the height is up to 2 feet. The male bird has *thondam* (throat hangs) of 3 inches. Sometimes Reddis domesticate this bird.

The above bird species have rather similar diet and feeds on fruits, grains, insects, decay, flies, ants, dead bodies, fish, other aquatic animals, etc. They live in thick bushes and on the trees in the forest. Most of the birds mate at the onset of rainy seasons.

Konda Reddis not only eat flesh of many birds, but also consider them as having immense medicinal properties. For instance, the eggs of the nocturnal bird, *Kamledu*, can cure jaundice. The shells of the eggs are burnt and the ash is applied to the eye for the prevention and curing of eye ailments. The meat of *Chambaru buraka,* is also used for the treatment of fits, paralysis, back pain, etc., although people are scared of this bird, as it hurts humans. It is also believed that in the nest of this bird, *sanjivini* (nectar) is found, which can make a dead person alive. *Thondam kodi* and turkey *kodi* are fowls, domesticated by the Reddis, as these can fight with snakes and eat them up. Of-course, both their meat is consumed by Reddis. Birds are also associated with good and bad omens. For instance, if *Annor gedda* comes into the village and cries near the village, there is a possibility of death. Like other animals, therefore, birds are integrated with the lifestyle of Konda Reddi.

Uses of Animal Parts for Human Healing

Konda Reddis may not have detailed knowledge of the diseases of the wild animals and their cure, but they know their medicinal value. Reddis often use animal body parts for healing human and livestock diseases.

The fat of the tiger and bear is used to heal diseases like joint pains (*Keedla noppi)* and also *Vatam* (swelling body parts). When they hunt the animal they take out the fat from the viscera. To extract fat from the skin they heat skin on a pan.

This fat is rubbed on the joints for relief. The meat of Udumu (*Varanus monitor*) is used to heal back pain (*Nadamu noppi*). The animal meat is cooked into curry and is eaten with rice. If the curry is cooked in the morning it is consumed in the evening and if it is cooked in the evening it is consumed in the morning. The meat of a porcupine (*Nedi pandi*) is very useful medicinally. According to Reddis, this animal eats all kinds of medicinal plants. Except the flesh of this animal, all other parts like heart, liver, intestine, etc. are said to be bitter. Reddis give the meat of the porcupine to women who have just delivered for a speedy recovery from anaemia and for curing of wounds. By eating the meat of the porcupine the new mother gets more energy. The meat of *Rushipitta* is used to cure joint pains. The meat of *Banda* is used to cure *Vatam*. *Rushipitta* and *Banda* meat is cooked into curry and is eaten with rice. Chicken is cooked by mixing some herbs and is used to heal joint pains. Reddis feed the meat of garden lizard (*Thonda*) for curing the carbuncle (*Rasapundu*). The meat of duck (*Bathu*) and cranes (*Konga*) is fed to patients who are generally very weak.

Knowledge of Insects

Konda Reddis classify insects on the basis of their mode of locomotion (flying, non-flying) colour, general characters, suitability of insects as food (edible and non-edible), their utility (harmful and harmless), their seasonal availability and their location etc. They call insects as *Purugulu,* flies as *Egalu,* ants as *Chemalu,* butterflies as *Gaga* and lizard as *Balli*. Konda Reddis do not worship any insects. According to them their ancestors also never worshipped any insects traditionally. Konda Reddis identify various insects found in the forest and in their habitat. The insects and their local names are: *Boddem Purugu, Garase Purugu, Aarudra Purugu, Etha Boddem Purugu, Bandare purugu, Gongali Purugu, Pacha Purugu, Uppu Purugu, Chedi Puru Purugu, Millimisika Purugulu, Chemalu, Surechimalu, Errachemalu, Puli Balli* (wild spider), *Uduku balli, Raballi, Penku purugu, Tummedda, Gommari Purugu* or *Kodi Pendlu, Rani Chema, Guntru chemalu, Pulla chimalu, Tene pillalu, Gaga* (butterfly), *Kundakanduru, Putta Kanduruega* and *Paccha Kanduru.*

Konda Reddis consider *Boddem Purugu* as edible and eat them by roasting. *Boddem* larvae are eaten by preparing curry, which is considered as a delicacy in the winter season. Insects are commonly found in the winter season on the *borassus palm* trees. These insects are very harmful for the palm trees as they feed on the inner pith (Muvvu) of the tree. This eventually leads to the death of the tree. According to Konda *Reddis* some tribes like *Basalu* eat *Etha boddem Purugu*. Reddis consider *Etha boddem Purugu* as harmful to the palm tree as it feeds on the inflorescence. *Reddis* say that *Gongali Purugu* is seen in the months of August and December, is a harmful insect as it feeds on the leaves and tender shoots of vegetable and fruits crops. *Chedipuru Purugu* feeds on the sap of *Nalla Chedi* fruit and is therefore named as *Chedi*. These insects are mostly seen in winter. If they crawl on the body, Reddis complain of itching sensation and boils on the skin. *Siri chemalu* and *Erra chemalu* are seen in the places where sweet things are stored. *Erra chemalu* and *Nalla chemalu* are mostly seen in the vicinity of mangoes trees. *Puli balli* is poisonous and their bite can sometimes be fatal. *Raballi* lizard lives in palm tree and drinks the juice of this tree. *Kanduru egga* feeds on honey and juice of the trees. *Reddis* in interior places eat these insects by roasting and also by sometimes preparing curries. It is said that *Penku purugu* spoils bamboo and other wooden parts of the roof. *Tummeda* insects are found on the human and animal excreta and feed on it. *Tene pillalu* are the larvae of the honeybees. These are found in the honeycomb (*Tene atta*). *Reddis* eat these larvae by roasting. *Konda Reddis* eat *Pulla chimalu* and *Rani chemalu* eggs by preparing curry. They lay eggs in the summer season and are mostly seen on the mango trees. According to *Reddis Guntru chemalu* are useful as they feed on other insects that suck the palm trees. *Gaga* insects feed on nectar of different flowers and prepare honey. *Konda Reddis* say that the poultry is infested with *Gommari purugu* and *Kodi pendlu* insects. To escape from the attack of these insects *Reddis* smoke their houses or they tie the leaves of *Billa Rotte* to the ceilings of their houses. *Millimisika purugulu* are seen mostly on the Tamarind trees and they look beautiful when they fly in dark nights. *Uppu purugu* live in the holes of the ground. *Pacha purugu* are harmful to crops like cotton.

Hunting Implements

Hunting is a part of the subsistence activity practiced by Reddi men only. Nowa days hunting (*Veta*) mostly takes place in summer whenever they find leisure time. They use bow and arrow (*Inti badda*) for shooting the animal, and sickle, axe and spear are used for cutting the game. Sometimes, hunting supplements the meat food of the Konda Reddis. Children learn hunting activity and techniques from their father, grandfather. People in the age group of 15 to 70 years go for hunting. Group hunting is commonly practiced in the community at the time of the village festivals like *Gangalamma* and *Bairudu Pandaga*.

Knowledge of Hunting and Types of Game Animals

Hunting is another subsistence activity that is invested with certain cultural importance, which encourages group activity and participation. Apart from its subsistence value, people think that going for hunting is also a kind of entertainment. People consider hunting as the major activity which profiles and projects masculinity. For Reddis summer months are meant for hunting wild animals. This is the pre-monsoon hot and dry period, when there is not much agricultural work for Reddis and they are in a mood for fun and recreation. Therefore both small and extensive hunting expeditions take place during this time. This does not mean that they do not hunt at other times. They set out for hunting after their mid day meals. When they set out to the forest and a game is noticed, brothers, kins, friends are alerted, and hunting takes place. Hunting is entirely the activity of men, and women are not allowed to take part. People believe that hunting in the surrounding forest is their birthright, and any restriction that keeps them away from hunting is never accepted and tolerated. Children are taught hunting techniques and the use of bow and arrow at the age of eight to ten years. The children, youth and elders from different groups go for hunting. Nearly 25-30 people go for hunting in bigger expeditions. Hunting of small games and birds is done individually. The real hunting group consists of all the youth and elders of the community. Every year hunting activity starts on the first day after the *Gangalamma Panduga*. This day is observed as an auspicious one on which all the males of a group have to inevitably participate in the hunting. The activity of hunting starts with some ritual performances. All the male members of the community from 8 to 60 years assemble either at the center or at times in the out-skirts of the village, at a place called *Kolatam Basa* where the village deities are worshipped. A few women apply turmeric, crimson paste and put *bindi* on individuals and pour water on all of them. Later, women throw cow dung and mud on men who are their cross cousins to symbolically drive them into the forest for hunting. Then all the men assemble in front of a chosen jackfruit tree (which is owned by the community) and each one is given a chance to shoot a particular jackfruit with arrows. Before shooting *pujari/goravadu* sacrifices a fowl in honour of *Gangalamma* (village deity) represented by a decorated pot under the tree and offers the blood by sprinkling it over the pot while chanting hymns. The game is started by any one in the village. Those who succeed are considered the best shooters who definitely go on to hunt animals. Then the group split into two small groups and moves in two different directions (opposite) around the hills. The first group is of youth especially those who are inexperienced in hunting, accompanied by instrument called *Gilaka* in order to drive away the animals by making sound and by shouting at the other group. The group is called *Sompu-Gumpu*. The other group is of efficient hunters and is equipped with bows and arrows. This group is called *Kapu-Gumpu*. On this particular day the groups together are bound to hunt at least one animal, although sometimes they come with empty hands to the village. Children below 12years locally called *Guntalu*, do not normally hunt. At the age of eight parents prepare small bows and arrows and give them it to the children to play. The children learn by observing the elders. At the time of hunting unskilled people go in the front and try to shoot the animal. The unskilled powerless people are locally called *Salladanam Chetulu* and skilled powerful people are called *Aggi chetulu*. First, the experienced people ask inexperienced people to chase the animal to their side, so that they can shoot the animal. The wounded animal, locally called *Jada,* runs bleeding after the shooting. And the drops of blood, which come from the animal's wound is locally called *Ella.* Hunting is considered easy in night, as they use battery powered torches and musical instruments to get the animal out of its hideout. When the hunters' light falls on the animal, the game cannot see the way to run. Secondly they use the musical instrument called *Gilaka,* which gives a louder sound, different from a foot sound to deceive and confuse the wild animals.

In the forest locations, where the game hides or often comes to drink water, some persons sit under a tree and watch. If a big game is noticed, they call other people. Every body carries with them their bows and arrows for hunting and most frequently they take their dogs with them to help chasing the game. When the game is noticed, the common practice is that about 15-20 people surround the area standing at different corners with their bows and arrows and scare the animal to surface out of its hide. If the game does not come out of the hide, the Reddis keep themselves ready with bows and arrows on three sides of the hiding bushes and a small group of 5-8 people chase the animal from the fourth side with loud shouting like *Puchhu, Ohoho, Vayyi,* etc. The role of the dogs at this time is very crucial. Dogs invariably enter the bushes and chase the already scared animal to come out of the hide. If the game is small, for instance, a rabbit, it is picked out from the bushes by the dogs. Otherwise, when the scared animal tries to escape towards the bushes, the alert hunters kill or wound the animal by arrows. Till the game comes out from the bushes, the hunters do not normally change their positions. If any one of the hunters, particularly the young among them, miss the arrow, he is reprimanded casually.

Reddis are very well aware of the behaviour of the games. For instance, they say that wild goats are normally tree climbers; *Adavi gorre* comes out to drink water alone to the water source; wild boars come out in groups for food, except the very young ones, who may be found singly due to some confusion. They also know that the hiding place of rabbits is thick bushes, where these may be found singly or in groups. *Duppi, Kanasu* live inside the holes of the rock, which can be hunt, only them they come out of the holes. *Gorragedde* is the animal, which mostly moves in groups. The right legs of the animal have white patches and the colour disappears, when the animal is dead. The big animals like wild boar, wild goat, deer etc. hide on the top of the hills where water is available. The hunters wait for the games exactly in these places. Sometimes, at the time of hunting when hunters make sounds to scare them, some small animals hide themselves inside the holes. In order to extricate these games from the holes, Reddis put fire inside the holes, forcing the animal to come out. Sometimes, however, due to the smoke emanating from the fire, the animal dies inside the hole, which is then extricated.

In the rainy season, the Reddis follow the footsteps of the animals for tracing them and hunting. They carry the hunted games, like the wild boar, by tying the legs with rope and inserting a bamboo stick between the two legs of the wild boar to be carried by two persons on their shoulders. If the hunted animal is too big for two to be carried back to the village, four people accomplish this task. Sometimes, for bigger animals like the deer, they make four holes right on the body of the animal and it is carried by eight persons.

Reddis view hunting as a means of subsistence. They participate in hunting expeditions and kill different animals. Reddis have a strong hunting preference for certain species, such as deer, sambar, wild boar, porcupine, rabbits, etc. But hunting is no more a source of bush meat, hides and ivory. In fact, many Reddis hunt wild animals once a year during *Gangalamma* festival. Previously, they used to hunt during the dry seasons in January, February and March when the animals are very weak and cannot get enough water to drink and grass to graze. Wild boar is used to hunt to sacrifice at the *Gangalamma* ritual. At present, antelope, boar, hyena, pig, small gazelle, tiger, monkey, colobus monkey, civet cat, wild cat, fox, *caoo*, hare, chipmunk, *boortee*, rabbit, tortoise, porcupine are common in the study sites. Even though many species of animals are conceptually recognized, Reddis use limited number of species in their diet. The following games are mostly found in the nearby forest and are preferred for hunt: wild goat (*Adavi gorre*), wild boar (*Adavi pandi*), wild goat of relatively small size (*Eetha gorre*), rabbit (*Chavulapilli*), deer (*Ginka* or *Karudeduppi*), *Kanasu, Nedipandi, Udumu, Nemali*, etc. These are bigger games. The small games include *Puttapilli, Pandupilli, Rasieluka, Vedurueluka, Peddamurada, Benthu, Adavieluka, Seemaamugeesa*, etc.

While hunting big and ferocious wild games like wild boars, special hunting techniques are followed, as there is every possibility of the game attacking the hunters. If a wild boar is noticed in the forest one of the hunters sits on the branch of the tree to keep a constant vigil on the movement of the game while message is sent to people to join him. For this kind of games, Reddis take more help from the dogs. Dogs are instructed to constantly chase the animal and make it tired before the actual hunting begins. When the game is tired enough and is not in a position to put up a tough fight with the dogs and hunters, arrows are sprayed from all sides to kill it. If any dog is considered not good enough to chase the game or to follow instructions given by the master, one of its ears or the tail is cut as a punishment and warning, so that it does not repeat the same in future.

After hunting the animal, Reddis take it to the nearest stream and wash the animal thoroughly with spring water. After cleaning the body, small games are put into fire in the forest itself. In case of a relatively larger animal, they separate the head, intestines and the legs after cleaning it in the spring water. Reddi men then cut the game with the help of sharp knives, such as, *Geethakatthi, Potekathi* and *Goddali*. The meat of the game is normally shared among all the participants equally irrespective of their role in the hunting. However, the hunter whose arrows pierce the body of the animal to kill it is entitled to get the special share of meat including the legs, the heart and an extra pound of flesh of the animal. The share of the first hunter is kept aside first. Then the remaining part of the animal is distributed equally among other participants of the expedition. Traditionally, if a *Duppi*, or *Kanasu* or an *Adavipandi* was hunted, the flesh of the game is usually divided into three equal parts. Out of those, 2 parts are distributed equally among the members of the hunting party. The remaining part of the meat is distributed to the whole village, however small the quantity might be. The village headman always gets an additional share of the meat because of his political position. But this practice is currently abandoned as the Reddis think that the hunting parties engaged in such a risky and hard work alone deserve their shares and not others. Reddis follow certain taboos with regard to the consumption of the forest meat. For instance, pregnant women and infants are forbidden to consume *Thaale mansam* (meat of *Thaale* animal). If by any chance the meat is consumed, the family has to purify itself by observing an expensive ritual called *Vetapanduga*.

Though Reddis get their staple food from their crops, they continue to rely on wild animals for food. The wild meat commonly called, as *Adavimansam* is very important in the livelihoods of Reddis. Hunting wild animals is a common practice among them mostly for food. Wildlife has always been utilized as a common property resource. Traditionally, hunting and trapping involved the use of local equipments such as wooden spears and ropes from vines, torchlight and the catch was mostly for household consumption. Individual hunter set some animal traps, and fence traps commonly used to increase chances of catch. They may also blend hunting with other activities, taking their weapon along to the field, ready for any opportunity that may arise. The Reddis are usually quite aware of the legal and administrative wildlife regulations, hunting a great variety of game year-round but preferring big and/or desirable animals, and those which can be captured by hand or with some simple tool. Hunting methods include tracking, shouting, humming, whistles or calls to attract animals. They also hunt a variety of birds, for them almost all wild species are clearly important. Hunting waterfowls, reptiles and other animals is also very important to them.

Reddis hunt to feed themselves. They generally share and eat the product of the hunt with the community. If the game is scare in the vicinity of the village, the most heavily frequented hunting grounds lie between five and eight kilometers from the village, implying that hunting expeditions were of more than a days' length in early days, but now they do not spend a lot of time these days. When hunting drops to unacceptable levels, there are three main options: (1) To move the settlement to a new location where the wildlife is abundant. (2) To go on hunting expeditions to more distance areas, perhaps rotating hunting grounds and gearing expeditions to hunting efficiency, so that areas where wildlife is scarce are visited less often, allowing the stock to recover. Both systems generate temporary resource recovery areas. (3) The third alternative is to abandon the traditional pattern, substituting meat from the domestic animals and pulse crop for bush meat, fish and so on. This is probably the case in most eventualities, and thus the need to manage and regulate hunting by the Reddis. Hunting is necessary but not absolutely essential for the Reddis. They share the kill with their family and sometime with the neighbours. But the acculturation is now leading to trade within and outside. In the Reddi settlements they don't devote themselves mainly or exclusively to hunting. But in some cases if they trap any animal in their fields and if there is excess meat, it is sold through personal communication. Some times if a group of people kill any large animal *Duppi, Kanasu, Adavipandi* etc., the meat is sold in the village and also sold outside the community. Bush meat is somewhat cheaper than that of domestic animals i.e. it actually depends on the hunted animal. Small-scale game sales are usually quite widespread in villages and hamlets of Reddis.

Although hunting wild animals and birds is very popular among the Reddis, this provides them only with food supplement. Trapping of animals is no less common. There are at least 5 different types of trapping, as observed in the village. Reddis find it easier to catch wild animals by trapping them while watching their standing crops at the *Podu* fields. This serves the dual purpose of crop protection as well as trapping an animal for a sumptuous meal. One of the trapping methods used is popularly called as *Budde*. For the operationalization of this method, 4-5 persons need to be involved, and normally it takes 3-4 days of preparatory time. First, a huge hole is dug right in the *Podu* field with a big net in it, made up of *Addanaara*. It is supposed to be a very strong net, the upper part of which is connected to a huge stone and wooden blocks are kept at the mouth of the *Budde* above the ground. The mouth of the hole is usually covered by tree branches and bushes to give a deceptive look that it is a part of the plain land. When the animal comes and unknowingly gets trapped in the net, the net shakes due to the weight of the animal. This results in the pouring of stone and wooden blocks into the hole, instantly locking the animal inside the hole, and sometimes, even killing it. Almost all types of animals visit the *Podu* fields to eat they damage the standing crops and normally get trapped in the hole by this method. Large sized animals including the wild boars, wild goats and wild pigs, etc. are caught by this method of trapping.

A second method for trapping animals is known as *Kudde*. It requires 2-3 persons and 2 days of preparatory time. Here too, a hole is dug for the animals to get trapped. About 20 bamboo or wooden sticks are firmly placed around the hole and another 20-30 sticks crisscross the earlier sticks covering the hole. The sticks are cleverly connected with large boulders and wooden blocks. A net made up of *Addanaara* is placed at the mouth of the hole. When the animal unconsciously passes through the hole, it is caught by the trap inside the hole, which is then filled with boulders and wooden blocks, trapping or killing the animal there. The third method of trapping is known as *Mettapadatti*. This a strong trap made with *Addanaara*. It takes a day's time to prepare for a skilled Reddi. The *Mettapadatti* has an opening gate, while the other sides are closed by bamboo splits. The gate is linked to a bamboo gate. When the animal gets inside the trap, the bamboo gate is automatically closed due to vibration, trapping the animal inside. It is the best method for trapping small animals, as large animals might break the gate open and escape. *Uchhu* is the fourth method of trapping the rabbits, which are the real threats to many standing crops in the field. *Uchhu* not only prevents rabbits from damaging the crops, but also is a useful method of catching them for meat. It is made up of strong synthetic threads, which Reddis buy from the weekly market for Rs. 30.00 per kilogram. The trap is designed in such a way that if any rabbit or other small animal enters into it, the animal is immediately trapped inside. It is usually placed in strategic places of the field, which are considered to be the regular pathways of rabbits and other small animals. *Adapepadpatti* is the fifth method of trapping the birds, which is the most effective method of trapping them. It is made up of half-split bamboos, separated by either a piece of wood or bamboo. Some grains are kept inside the half-split bamboos to attract the birds. When the birds sit on the bamboo split to eat the grains, the separator gets disturbed closing the bamboo split and trapping the bird. There are many other methods of trapping animals and birds in which Reddis are adept right from their childhood.

The key groups and species present a broad diversity of biological traits. Some species or groups exhibit fairly marked sexual dimorphism, enabling selective harvesting by sex. Most of the key groups and species are forest-dwellers, aquatic, of mixed or secondary habitats, or found in wetlands etc. Except for the big carnivores, forest-dwellers, and some of the aquatic species, they are usually, partially or wholly frugivores. This dietary preference is also shared by quite a few birds and mammals. As a result, the diversity of fruit, and the phenological aspects of fruit production and availability, are crucial factors for wildlife and deserve top priority in any review of wildlife habitats.

The reproductive periodicity of the key species is only partially known, apparently varying greatly from one species to the other and from one place to the next. Seasonal reproductions in some animals are clearly defined. They lay their eggs during the dry season and the clutches hatch with the onset of the rains. Most birds nest during the rainy season but the length of the breeding season often varies from one place to the next. Wild animals breed on a seasonal basis in the spring or at the onset of summer. A whole series of animals exhibit peak sexual activity during the wet season with birth peaking during the transitional period preceding the dry season. Quite a number of these species also breed at other times of the year as well. A considerable number of species apparently breed year-round. This includes some mammals. Some tend to breed more during a specific season. The interactions between wildlife populations, habitat and people are complex ones involving many aspects of wildlife biology. Some, however, are of more immediate utility for management than others.

The Reddi knowledge of the behavior of wildlife may be as detailed as the hunter/gatherers. Their knowledge of wildlife behavior includes much attention to words the details of behavior and ecology of practically every animal. Animals and insects are used for food, tools and indicators of ecological dynamics. However, their consumption as food among Reddi is often accompanied by taboos that restrict the consumption or killing of certain species. For example, among the *Reddis* some animals are not eaten because of supernatural powers or because of ancestral taboos. Other wildlife rabbit, birds, insects, eggs of honey are eaten by all. The diversity in the use of animals is usually less than in plants. The knowledge of diseases and their cure is quite high among *Reddi* population, who have to depend on local means to mitigate the effects of diseases.

6

Local Knowledge and Management of Natural Resources

The traditional utilization by the indigenous people of the biologically diverse resources in the region not only reflects a diverse resource-use pattern but also a wise way of maintaining the region's biological diversity. The conservation of resources over a long time depends on community management. In the course of time, many forest tree species are integrated into the agriculture system and useful forest species that regenerate spontaneously in the garden, individual forests, and in *Podu*, fallow areas are important for sustainable forest management. There has been great awareness about the variability of crops and plants, including wild relatives of crops. In fact, this region is considered to be the primary and secondary source of many plants found in the country today. This assemblage of plants and their wild relatives presents a valuable wealth of genetic variation, which could be utilized for current plant improvement programme, to cater to future needs and meet new demands. Natural resource management systems are localized systems, which form the basis for decision-making by the rural people. Most of the land-based production in the region is based on local knowledge systems. These systems are not only of value to cultures from which they evolve but also to scientists and planners striving to improve conditions in other rural societies.

The indigenous communities play an important role in generating knowledge based on understanding of their environment, devising mechanisms to conserve and sustain their natural resources and establishing community-based organization that serve as forum for identifying problem and dealing with them through local level experimentation, innovation and exchange of information with other societies (Warren, 1992). Observation of nature and through elementary reasoning based on such observation, the communities have accumulated a store of working knowledge concerning the effects of certain elementary mechanical processes, the apparent movements and functions of some of the heavenly bodies, the habits and haunts of animals and birds, the properties of plants, fruits and flowers, barn and roots, the nature and qualities of different kinds of soil and the variations of weather (Roy 1928). Reddi community in the study village possesses highly detailed and richly complex information about flora, fauna and land. They recognize both natural and super natural forces and agencies shaping human destiny and seeks to utilize them for their benefits according to their needs.

Indigenous knowledge is historically constituted (emic) knowledge instrumental in the long-term adaptation of human groups to the biophysical environment (Purecell, 1998). Indigenous knowledge of the village community can be divided into two different domain, one that of cognitive and other that technical relating to livelihood issues. The cognitive knowledge is understood by the worldview of the community about the society and its relationship with its outside world. The technical knowledge is information relating to understanding generated in past about the environment and exhibited through the practical application of it in fulfilling their livelihood activities.

Discussions in this chapter have been made about the local knowledge system and natural resource management of Konda Reddis.

Local Knowledge of Drought Management

Reddis say that the drought, floods and famines, arise due to drastic changes in the climate (*Vathavaranam*) and destroy the forest and the environment. According to Reddis, droughts are of two types based on rainfall. First one *Aanavriushti* means the drought caused due to lack of rainfall for which the crops are affected. Second one is *Aathivruishti* caused due to heavy rains due to which there is no crop production. *Musabu* is the local term for the floods and it never occurred in the study village. *Reddis* attribute droughts to both supernatural forces and natural reasons. Some of the elders say that droughts occur when sin grows among the people. They say, god punishes the sinners by causing droughts. In the study village Reddis faced drought two times earlier, first time fifty years ago and again around thirty years ago. Both droughts were very severe and there was acute scarcity of food due to the failure of monsoon and crop production. It is common in these areas on an average a drought occurs once in 5-7 years. At the time of drought, people do not have any crops to eat and the village people eat the following food: tamarind seeds (*Chinta pikkela Ambali*), gruel (*Mamidi tenkayala Ambali*), and chapati (*Mamidi tenkayala rottelu*), made from the flour of mango kernels. Broken rice (*Nukkalu*), bamboo rice gruel (*Katukulu Java*), *Gante nukalu*, and *Allepandlu* in rainy season. *Peddapala pandlu* and *Tati pandlu* in summer season, wild tubers (*Adavi dumpalu)* and wild leaf vegetables (*Akukuralu*) and fruits from the forests. At the time of drought no other crops are grown except *Gante* because this crop does not need water. And they also buy broken rice (*Nukalu*) from the shops. And some of the other crops like *Gante, Sama, Korra, Bonthalu, Allu, Kandi* are got from labour work. During the last drought many *Reddis* went to Bhadrachalam in search of employment as wage labour.

During the droughts the only source of water is the underground water reserve, which flows as a stream (*Uta*) throughout the year. And the village people used to manage water for drinking from this stream. Both the villagers and cattle are dependent on this stream for drinking water. In summer when the *Reddis* go for hunting, to quench their thirst, they find out a wet place (*Chammaga Una Uta*) and then dig hole up to a foot deep with hands. Water gets collected in that pit after a while, which *Reddis* purify with *Dopa* (which is prepared by *Tati Aku* or *Moddugu Aku, Tumika Aku, Adda Aku*) and drink. They call this water as Chalama Neeru. Animals also dig such pits called *Gottulu* or *Guntalu* in summer to quench their thirst. Besides these, *Chamma Uta* pits are dug out on the banks of canals, rivers, streams, etc. If *Reddis* do not find water in any source when they are out for hunting in deep jungles, they take the watery sap from two kinds of tendrils: *Buruda teega* and *Yerrachikum*. These help them to wet their throats. They say that consuming more sap induces itching in the throat.

During severe drought conditions, villagers celebrate *Kappa Panduga*, a festival for the rainfall. To celebrate this festival every household contributes some money and provisions. On this festival village people worship a frog that is *Kappa devudu*. First they bring a frog from the stream (*Uta*) or canal (*Kaluva*) and apply turmeric on the body of the frog in front of the Goddess *Ganagalamma*. After this they shift the frog into a *Kavedi* and take a procession in the whole village and finally leave it in the water of the stream. While leaving it finally in water if the frog croaks, then it is an indication for them that rains will follow. Once the frog croaks, the village celebrates by playing with colours and sprinkling water on one another to symbolize the rains drenching them. After this they sacrifice a chicken or goat in honour of Goddess *Gangalamma* and all dine together. *Reddis* are confident that rains follow the croaks of the frog.

The drought spell was blamed for the decrease in the use and consumption of some indigenous crops. The drought has contributed to depletion of the indigenous seeds and farmers are now using hybrid seeds, some of which are not suitable for the preparation of traditional foods. Certain species of indigenous trees did not survive the drought, hence they are no longer available to the farmer for various purposes, as fruits and for medicine.

Conservation of Sacred Groves: Local Biodiversity Zones

In Reddi religion, the sacred groves i.e. *Gangalamma* is the space allocated for the village deity and all the activities related to them are carried out as a collective phenomenon. Religious beliefs and ritual practices are complete with the tree worship because many of the tree species are mythically connected with various Gods and Goddesses to secure a formal religious status for them. Besides a number of plant parts are of economic, societal and medicinal values for the Reddis. Certain plant species and their parts play a significant role in the ritual and life of the village people. *Reddis* consider number of plant species as sacred.

They use plant and plant parts in the ritual celebration. *Pala chettu* (*Mimusops hexandra*) is a tree, which found to be present in each grove as this is the dominant tree species growing in this region. The sacred site in the grove is mainly located under one such *Pala* tree. Different parts of this tree are used in the ritual of *Gangalamma* festival. The branches are used for decorating the place of ritual. Leaves are used for making plates and cups, which are used in offering made to the deities and later on for distribution of sanctified food. If it is not available they use the leaves of *Adda chettu.* The jute *Adda chettu* is also elaborately used by the villagers in the construction of houses, sheds and boundaries of the courtyard. The seeds of the fruits are collected for consumtion and sold to the GCC department. The branch of Somi tree is used for making idol of *Gangalamma*. *Rella (Cassia fistula)* and *Nallamaddi*, *Yegesi* trees are used in celebration of *Gangalamma* festival. The tree wood is used for making cradle (*Vuyala*) of *Gangalamma*. The flowers of *Pala chettu* and *Rella chettu* are used for worshiping *Gangalamma*. Similarly the fruit seeds of Amodam tree are considered to be sacred. They extract oil from the seeds for light at *Gangalamma* deity. The leaves of *Vepa* tree are used in warding away of evil eye.

The tree Neredu (*Syzygium cuminii*) considered to be sacred and different part of this tree is used in celebration of marriage ritual. At the time of marriage or before construction of house, two days before they first dig the pillar of *Neredu* at North side. The twig of this tree is used in warding away evil eye. The pillar of *Neredu* is tied with the leaves of neredu, mamidi and raavi along with turmeric and rice and 1-2 rupees, which signs as *Murthum rata* (Inaugural pillar) on the eve of auspicious occasion. The leaves of the tree are considered to be auspicious and symbolize prosperity, fertility and productivity. The tree log is used for making elevated platform (*Peetalu*) for the bride and bridegroom to sit at the time of marriage. To seek good life for married couple it is necessary to plant a *Neredu* branch at the corner of the house or marriage pandalam.

Reddis celebrate many festivals and rituals with the leaves of mango (Mangifera indica), which stands for purity, sanctity and prosperity. The leaves of mango are used to decorate the entrance of the hut doors. The bark of mango and raw turmeric are made into paste, mixed with water and sprinkled on the bride and bridegroom at the time of marriage, to get rid of all misfortunes and evil eyes. *Pacha pasapu (Curcuma longa)* and *Kunkuma* (Vermilion) has high religious value for the *Reddis*. At the time of every ritual and festival without turmeric and vermilion they don't perform any ritual and festival. They grind turmeric and apply it to the bride and bridegroom to give marriage bath and it is considered as purity. In the marriage ritual at the time of tying *mangalasutram* elder people sprinkle *talambralu* (Rice and turmeric mixed powder) as a kind of blessing to get rid of misfortune. They also apply to God and Goddess at the time of ritual. After applying the turmeric they put vermilion dot. Pedda tungagaddi (*Cyperus rotundus*) the root of the grass is used in *gandam* and the paste is applied on the body of bridegroom at the time of marriage as good perfumer.

The leaves of *Pala, Neredu* and mango and turmeric powder, mixed in pot full of water is used to sprinkle at the place of death and at the burial place to make them free from pollution. They offer grains on the leaves of *Pala* and neredu trees in honour of the departed soul. If the pregnant women die they stroke *Udike manu* tree wood nails to the hands and legs and also apply *chedi* to the eyes and burn to the body to free from pollution.

In the study village, there is tradition of worshipping and preserving the following wild tree species. The most honoured among the trees are the Burugu chettu (*Bombax ceiba*), Vepa chettu (*Azardirachta indica*), which are associated with village deity *Gangalamma*, Tenka rai (*Ficus sps*) is a sacred grove associated with forest God *Bairudu*, Chinta chettu (*Tamarindus indica*) is associated with *Chintalamma* Goddess. *Reddis* say that *Chintalamma* Goddess resides under this tree along with many other Gods and Goddess. At the time of marriage both bride and bridgegroom will visit this tree to seek the benediction of Gods and Goddesses. Raavi chettu (*Ficus religiosa*) is associated with ancestors' spirits. The above tree species, under which are the abodes of different Gods and Goddess, are associated with various rituals. Thus the trees protect the *Reddis* from the attack of evil spirits and misfortunes. Velaga (*Feronia elephantum*), Marri chettu (*Ficus tuberculate*), Neredu (*Myrtus cyneimum*), Maredu (*Aegle marmelous*) are some other species, which are considered to be sacred. Therefore the trees symbolize sacredness and have immense religious and magical power, for which it is worshiped by the *Konda Reddis*. When a Goddess or a God is found under a tree or a rock they treat the rock or tree as a sacred. Climbing on the sacred grove, or cutting the sacred trees, or selling or buying the sacred groves is strictly prohibited, as all these activities are considered as sins. *Reddis* say that if any body cuts the sacred grove, the person or any other member of their household or their domestic animals will definitely die. Sometimes the offender will contract some dreadful disease and he will be ostracized from the village. They believe that if the sanctity of these groves is disturbed, the villagers collectively will incur the wrath of the Gods and Goddesses. Before the festival starts the villagers arrange a meeting and discuss about the arrangements to be made for the festival. All the villagers contribute money to buy the sacrificial animals. On the festival day, they clean the sacred grove, smear the place with cow dung and decorate with lime stone powder (*Muggu*). After this, *Goravadu* (Pujari) performs the rituals to God and Goddess and an animal are sacrificed on the occasions. Plant species used in rituals and other ceremonies by the *Konda Reddis* are presented in the Appendix 19 followed by the local name and the name in English.

According to *Reddis* of the study village, Goddess *Gangalamma* was born under the *Burugu* tree, and hence is being worshipped from the time of their ancestors, dating back to a few centuries. This grove is located on the hills called *Bodi Konda*, which is about 2 km on the eastern side of the village in the forests. According to the elders the earlier settlements were near the *Burugu chettu* and it was known a *Patha Padu*. Later the *Reddis* moved out to the planes and settled at the present site and named it as *Utalapalem* after the stream. While changing their residence, they shifted the goddess Gangalamma to the present village and set it up facing east under a Neem tree (*Vyapa chettu*), situated at the center of the village. This Neem tree, they say it is about a hundred years old. *Reddis* give the reason for shifting the deity is that *Gangalamma* is a village deity, and she should be near to their settlement, so that she could protect the village. Though they shifted the stone symbolizing *Gangalamma*, they initiate the rituals from the old sacred grove *Burugu chettu*.

They assign specific functions and a place to each category of deity, both in their cosmology as well as in their village ecosystem. It was primarily because the Reddis way of life interacts with the local environment, the fauna and flora, the rocks, stones, pathways, fields, rivers, streams, forest as well as the people of the region in a closely interconnected way. Each sphere of life flows into other almost continuously. There is an underlying logic, but it is determined in a reinterpreted manner. It changes from time to time from event to event and from interaction to interaction.

In the hierarchy of deities *Gangalamma* is the supreme deity of the *Reddis* religion. They believe and consider her as the creator of the universe and the humans. *Gangalamma* is benevolent and has given the management of the universe, into to the control of various tutelary divinities and spirits or devils that the *Reddis* must propitiate. A sacrifice of country chicken and goat is usually made in the name of *Gangalamma*, whenever other spirits or deities of the village are being propitiated or worshipped especially during the principal *Reddis* festival of *Gangalamma*, which is held in the *Gangalamma* sacred grove. The village chief deity *Gangalamma* is invoked at the time of severe difficulties.

Ancestral spirits (*Parentalu*) are in the second hierarchy. *Reddis* are very close to ancestral spirits. In the community every person becomes a spirits after death. They use to bury the dead body. Except for the annual ceremony, there is no special religious festival on which offering or sacrifices are made solely to the ancestor-spirits. These spirits are involved at every feast and on every suitable occasion, their living relations make offerings to them. The ancestors spirits are usually remembered when one is seriously ill. A vow is taken to offer a chicken on recovery.

In the third layer hierarchy are deities and spirits of the village, who are regarded as Gods like *Bairudu*, Goddess like *Chintalamma* and other as *Bhutalu* and *Daiyalu* or *Chedimundalu* ghosts belong to this grade. The God and Goddess live in sacred groves spread around the villages. Here all the natural resources like hills; water bodies, etc. are assigned with a deity and are propitiated regularly on an annual basis or at a fixed interval of times as desired by the village community.

In the fourth layer hierarchy are certain spirits like *Gandhari* they are special spirits of women, and the spirits of hunting *Adavi rajulu*, who are worshipped by the young men. For the spirits of hunting they had to make periodical sacrifice and worship to these spirits as a means of propitiation.

In the fifth layer hierarchy are the spirits or mysterious powers residing in certain objects or symbols and the spirits of the persons met with un-natural deaths, apart from the evil eye, evil mouth, etc.

The village deities in *Reddis* religion are placed both in vertical and horizontal plane on the basis of function assigned to them by the society. As seen in the classification of deities in various ranked categories, with the specific ritual process prescribed for the community, the *Reddis* have provided different role to different deities/spirits.

The *Reddi* community seeks to ensure safety and prosperity to the village community as a whole at each stage in the annual cycle of its economic pursuits of the village such as agriculture, food gathering, hunting and fishing. All the activities associated with the economic pursuit of the villages are celebrated collectively. The Reddis celebrate its rejoicing and its sense of renewed social solidarity with feasting, drinking and dancing. Therefore there are series of festivals associated with various livelihood activities of the village system spread through out the annual economic cycle of the village community.

Gangalamma is the supreme Goddess and has power to control all the other Gods and spirits in the *Reddis* community. While worshiping the God and Goddess, basically faces east direction, which is the direction of rising sun. They do have grove or place allotted for *Gangalamma* in the village ecosystem and on every important festival occasion they offer sacrifice chicken or goat in the name of *Gangalamma.*

Gangalamma's control over the other spirits is evident from the arrangement and rituals, at the principal festival known as *Gangalamma Panduga.* The year starts by worshipping and sacrificing is offered to her, and later on to other spirits. This arrangement is meant to symbolize their belief that, whenever necessary *Gangalamma* controls the other spirits and sees that they behave properly. The *Reddis* believe that *Gangalamma* punishes offences against customary morality.

Gangalamma is the residing place of prominent deity named in the village, a female deity. She is considered to be a benevolent deity, who looks after the village production system. The grove provides productive and benevolent benefits to the community. This signifies healthy relationship developed among the human, nature and supernatural power for productive and benevolent functions. The attributes and functions of *Gangalamma* clearly indicate that she is in origin a nature deity, representing the earth, water or spirits of vegetation and symbolized as mother earth. The deity seated under an old *Burugu* tree located in the grove. A stone is placed under the tree to symbolize the deity. It is located in the northern direction of the village. The village community celebrates its main festival of the village *Gangalamma Panduga* in this grove of the village. A part from that the villagers perform other rituals and celebrations relating to village agriculture and social system in this grove. Goravadu of the village perform the rituals pertaining to *Gangalamma*. There is no restriction or taboo to enter or participate in *Gangalamma* on the occasion of the ritual. Country chicken is offered every year, in all the village festivals performed by the *Goravadu*.

According to *Reddis* of the study village this festival is first clebrated in Yelshwaram Mandal in the East Godavari region. There is a main *Gangalamma* temple and Konda Reddis of all other regions start celebrating in their respective villages after the festival is started in this temple. Every year an elaborate festival (*Gangalamma*) is celebrated in honour of the Goddess *Ganagalamma*. This is one of the important festivals celebrated by *Reddis* in the East Godavari region. This is an annual festival, the celebration of the festival is held for three days, which is celebrated elaborately since the whole village has by this time complete their agricultural activities. The initiation and culmination of the festival rituals is done at this sacred grove. *Gangalamma* is a village deity and *Reddis* believe that if she is pleased she will protect them and their livestock from contagious diseases, droughts, and from wild animals. The celebration of this festival is for three days in the summer in the month of May, after completion of all their agricultural activities. During the elaborate rituals, they express their gratitude and pray to the village deity, *Gangalamma* for undisrupted agricultural activities and sufficient harvest. In sense this festival can be noted as the annual harvest festival where all the harvested grains are ablated to *Gangalamma*. The village council, *Kulam Kattu* headed by the *Pujari* or *Goravadu* decides the day on which the festival should start. Which preferably falls on a Monday, the ritual starts in the morning. *Kulam Kattu* along with villagers decides about the probable expenditure and the pooling of money. The *bandroth* (messenger) gives the information about the date of *Gangalamma* to each house in the village. After fixing the date and day of celebration for *Gangalamma* festival, a fixed amount of money is collected from all the households. In this they exclude the houses of Christians. Sometime instead of money, a fixed amount of rice i.e. one *Kuncham* (four kilograms) is collected from each household. The grains collected are sold in the market. The price realized by selling the grains in the market is used for purchasing the material required for the ritual of *Gangalamma*. The material consist of four to five earthen pots, coconuts, *Kunkuma* (red vermilion), turmeric, lemons, winnowing fan, knife, thread, lungi, clothes for *Goravadu*, fruits and cock. If goat is also to be sacrificed, then money is collected for the same and animals are purchased as per specification of *Goravadu*.

The *Gangalamma* festival is of two types. The first one is *Potu* (male) *Gangalamma* festival, where they sacrifice he goat on this occasion. Another one is *Aada* (female) *Gangalamma* festival, where this festival celebrates with sacrificing chickens except goat. These two festivals are celebrated in alternative years in the month of May. First, the celebration of festival starts from *Yeleshwaram* where *Gangalamma* temple is resided. After celebrating the festival in *Yeleshwaram*, rest of the villages starts celebrating this festival. In the study village they start this festival on Monday and finish on next Wednesday. Every day in the evening *Goravadu* prepare *Gangalamma* from his house and send to all the other houses in the village. They make *Gangalamma* to one girl whose age should be 8-10 years, where the girl of the village keeps fast and worship to God. This year they made *Gangalamma* to 8-year-old girl whose name is Bachala laxmi.

The deity is brought from its abode and kept in the village for three days. Both the festivals of *Gangalamma* are celebrated similarly. The villagers start the festival in joyous mood by performing dance by male is *Kolata* and by females *Nandalatta* by playing *Baja* or *Dolalu* (musical instrument) on every night. On the day of *Gangalamma* festival *Goravadu* gets up early in the morning and takes bath, after which he returns to his house and enters into a room where the sacred pot and knife are kept. These two are used in the rituals of *Gangalamma* festival. He worships them and requests the deity of *Gangalamma* to sit on the new pot. The *Goravadu* takes a new pot and paints it with white lime and decorates it with *Pala Puvvu* (flowers) and keeps in it his house. After this they choose a girl between six to ten years, who is given a ceremonial bath and decorated with turmeric powder and vermilion (*Kunkuma*) like *Gangalamma*. Later she is taken to the house of *Goravadu* with the decorated pot on her head, marking the beginning of the procession. At the time of worship, a person will go to the Gangalamma place early, and cleans the place where rituals are to be conducted. Each and every day till third day *Goravadu* (Pujari) has to prepare *Gangalamma* from his house by washing her feet and worshiping in the evening and send to every house for worshiping to the Goddess. The girl and the procession will go to each and every house in the village. Since the girl personifies *Gangalamma,* in every household her feet are washed with water and then turmeric is applied, vermilion and cooked rice is given to her. Every day the procession starts at 4:30 am. Like this, the girl goes every day to each and every house up to three days. The cooked rice, which is given by every household, is called *Baliannum,* where the *Prasad* banana is mixed with cooked rice and is distribute among all the people. Since it is believed that by eating this *Prasad* they don't get any eye diseases (*Kalakkandlu*). Meanwhile the Goravadu goes to each and every household to collect a small handful of paddy. This is offered to the *Gangalamma* and later the sanctified rice is distributed to head of every household of the village to mix with the seeds kept apart for sowing in the forthcoming agriculture season, so that there is surplus production of the grains in the coming year.

At the time of worship to *Gangalamma,* first they apply turmeric and red vermilion to the pot and only vermilion to the girl. On the second day of the festival in every house in front of their house they erect a small pandal *(Pandiri*) with *Pala chettu* and take a small water container *Chambu* or pot and decorate with *Pala* flowers and keep inside the pandal. They treat the pot or *Chambu* as *Gangalamma* and sacrifice the chicken in front of the pandal, which they locally called

Dhodiwarulu. After this they cook the chicken and rice, which was collected from each house, and the cooked food is distributed among the people. On Sunday they bring *Maddi karra* or *Yegesi karra* to dig the pillars and prepare a swing (*Vuyala*) for the *Gangalamma*. All the villagers, especially women, swing on it. In these three days every men and women dance and sings in the nights. Men play *Kolata* and women play *Nandalatta*. On Tuesday, they make *Gangalamma* to sit in the cradle after that they first go to *Musif* house and later *Bandroth* to receive the gift for *Gangalamma*. Both of them present a blouse piece, break the coconut and give bananas to the *Gangalamma*.

On Monday i.e. first day of the festival *Goravadu* do the same thing and on the last day they take it to the house of the *Munsif*. On the last day they take procession of the village people by dancing to take *Gangalamma* into the forest where her actual place exists under *Burugu Chettu*. At this place they sacrifice he goat if it is *Potu Gangalamma* festival from the whole village. Sometimes they individually offer goat and chicken. After this all the people come back to the village, only *Pujari* and some people stay back to bring the sacrificed animal and *Gangalamma* back into the village. On the day of *Adda Gangalamma* festival they can sacrifice chicken also. Personally some families offer chicken to *Gangalamma* for their better crop.

For the rituals and ceremonies *Goravadu* prepare castor oil (*Aamodam Nune*). The preparation of castor oil takes place in a new pot strictly meant for this purpose. The castor oil is used to light a lamp (*Dibbi*) in front of the *Gangalamma* idol. In the evening of the second consecutive Sunday the *Goravadu* goes to the *Gangalamma* who is represented by wood idol situated under a tree at the spot of *Idelu* (*Idelu Stalam*) accompanied by *Baja* (musical instrument), performs a ritual offering a fowl, while chanting hymns and sprinkles blood on the idol and also in all directions. A small girl 8-10 years old carries the idol pot in a procession through the streets of the village accompanied by drums and trumpet.

First, they carry the idol from the *Goravadu*'s house where turmeric and castor oil is applied on the idol and the same girl takes the deity to the stream and cleans it in the water by applying *Sudda* and *Jaji Kai* powder. This process is called *Vanne pettadam* in local dialect. Later, the deity is carried back to the village and kept on an elevated platform on the eastern side of the village under a Neem tree, which is called *Kolatam Basa* that is situated between the village and the *Idelu* spot. All this process takes one full day that is on Sunday a day before the festival day. On Monday evening when the festival starts whole village assemble at the Neem tree to the accompaniment of the *Baja*, the *Goravadu* decorates the deity with new pot which is filled with fresh water and is placed beside the deity with *Paala pulu, Sampena Pulu, Rella Pulu* (flowers) and other flowers available during the season. The *Goravadu* also applies turmeric (*Pasupu*) and Vermillion (*Kunkuma*) powder. Later, the decorated deity and the pot is carried on the head of the girl and she goes to each and every household accompanied by the *Dollagallu* (musicians) playing *Baja* and with some children. At every household, the Maiden of the house applies turmeric and vermilion powder and offers *Dhupam* (fumes of incense). She also offers some rice and prostrates to the deity.

The practice of carrying the deity and the pot to each household takes place every day. Every time, the deity is lifted from *Gorsavadu's* house and kept back at its fixed place throughout the three days, during the night's men and women as well as children's dance to the rhythm of *Dolalu*. Before the final day, in the night they take the procession of *Gangalamma* to the *Munsif, Bantroth* and *Goravadu's* house to present the blouse piece to *Gangalamma*. On the final day, that is the third day of the festival which falls on Wednesday, the girl carries the idol of the deity to the *Munsif* house for *Vannepettadam* and the *Goravadu* decorates it with flowers, turmeric and vermilion powders as it was done on the first day of the festival and the *Gangalamma* is taken to her original place. On reaching the grove, they again clean the place near the sacred site for worship and sacrifice. Paddy is kept to feed the sacrificial cocks before sacrificing the cocks they are washed and made to eat the paddy kept on the ground. If the cock eats the paddy it is believed that deity is ready to accept the sacrifice. The head of all the fowls are not wholly cut but partly severed from the neck. If the goat is to be offered to the deity, then sacrifice is performed with the sacred knife possessed by the *Goravadu*. The *Goravadu* himself sacrifices the goat, some people assist him in sacrificing rituals. Before the animal is sacrificed vermilion and turmeric is applied to the sacred grove and coconut is offered to it. *Pala* and *Rella* flowers are offered to the village deity.

Goravadu performs rituals to *Gangalamma* along with male and female members of the villagers, where all the villagers come to *Gangalamma* bringing chickens and coconuts. Then the *Goravadu* places all varieties of food grains that are harvested from the field in front of the deity keeping each grain in an *Adda* leaf bowl to offer to the deity. He performs the rituals by tying unbleached thread *(Dharam)* to the grove in three or five rounds. Later while chanting the hymns he feeds all the varieties of the grains to the chicken and the goat and sacrifices them. He sprinkles the blood on the idol pot and also on the grains that are kept in *Adda* bowls by saying *"Alandu, Balandu, Podu, Meraka, Polam, Kachikapadu"*. Which means protect elders, children's, fields and the cattle of the village every year. Then he mixes all the grains in a bamboo basket which is meant for that purpose, accompanied by other elders and *Dolalu* he then goes to the four directions of the village and buries some portion of the grains in each pit saying *"Agralu"* later in the night, the whole villagers actively participate in dancing, playing *Dolalu* and playing games. In the morning of the third day, the *Goravadu* places the idol on its original spot accompanied by the villagers while playing *Dolalu* and sacrifice chicken and prostrates to the deity. After keeping the idol in its original place where as the decorated pot is taken to the *Goravadu's* house. The sacrificed goat is skinned and cooked with rice and is shared to every household among the villagers. The *Goravadu* takes up a small amount of cooked food and roasted meat and offers the same to the deities and spirits. The house made liquor and toddy plays an important role through out the celebration of *Gangalamma* festival. Few pots of liquor and toddy are prepared with utmost cleanness, which is offered to the deities and spirits and later on consumed by all villagers.

Gangalamma festival marks the end of the festivals associated with agriculture. After celebrating this festival, they start agricultural activities. This festival is celebrated in each and every village in the region, in their respective sacred grove. This festival is celebrated as symbolizing the productivity aspect of the prosperity on earth by good agriculture crops. It is also a thanks giving rituals for the previous year as well a welcoming of the new agriculture season. Until this festival is celebrated in the village no one gathers or use new fruits or starts agricultural activities. After the celebration of *Gangalamma* is over, the villagers start preparation of fields for forth-coming agriculture season.

After celebration of *Ganagalamma* festival, hunting festival is celebrated. There is one festival associated with the hunting activities of the village system that is *Veta Panduga*. In this they get *Panasa pandu* (jackfruit), assume the fruit as an animal and collectively shoot with bow and arrow if any person shoots first he will lead the team. They go collectively to the surrounding forest and do hunting. It is more associated with agriculture well being. If there is failure in hunt it is considered that forth-coming agriculture season will be bad for the village community. Therefore, all attempts are made to see that there is no failure in the hunting expedition.

On the occasion of festival the entire men of the village go for communal hunting. If anybody successfully kills an animal, he is lifted by his fellow hunters and is brought till the *Pandiri* (pandal). Near the pandal they dig a pit and pour some water into that and prepare muddy water. They play tug of war with the carcass of the hunt with men at one end and women at the other. The winning team pours mud water on the losing team. The hunting continues for three days. On the fourth day, they leave the *Gangalamma* at old place at the sacred grove of *Burugu Chettu* and bring the pot back and put it under the Neem tree. The tradition of maintaining the sacred groves is remained and enacted through community festivals.

Reddis also observe *Rajulba Pandaga* in the month of February every year in hounour of the forest god for protection from wild animals and evil spirits during their quest for food and games from the forest. They also celebrate *Dari Padaga* (to propitiate forest path) and *Veta Pandaga* (hunting festival) in the month of January every year to maintain their relationship with the forest and forest God. They also observe *Gaddi Pandaga* (Grass festival) to ensure sufficient fodder availability for livestock and for thatching roofs.

The villagers celebrate the ceremony of some trees like mango, *Thati chettu, Jeelugu* on the occasion of their first eating. They consider that unless this ceremony is conducted the fruits are not suitable for usage. After the conduction of first eating ceremony, the fruits of these are consumed.

Reddis during the month of April celebrate *mamidi panduga* of the first eating of mango. They celebrate this festival when mango trees are full with unripe fruits with green colour. The village priest *Goravadu* who decides an auspicious day for the ceremony, which preferably falls on a Monday, performs this festival ritual. On the day of ceremony people get up in the early morning collect five mangoes where as two of them offered to ancestors at the northwest corner of the house. Three mangoes tied at the entrance of the residence. Later in the day for the worship of tree they offer some slices of the raw fruit to the mango tree.

Konda Reddis worship the Jeelugu Chettu (*Caryota urens*) and Thati chettu (*Borassus palm*) where the occasion is commonly known as *Chettu panduga*. The toddy of the above trees considered being the most favoured drink for the *Reddis*. The trees take nearly 20-25 years to reach maturity, at this age it starts secreting toddy and it will continue secretion up to 20 years. Reddis extract toddy through out the year but the secretion of toddy is more from October to June. Before the extraction of toddy begins, the trees are worshipped with the expectation of abundance of toddy for its entire life. Any *Reddi* man of the community, who first claims the ownership of the toddy tree, is known as its ritually authorized toddy collector locally known as *Geetari*. On the occasion of the tree ceremony, on an auspicious day the *Geetari* or *Goravadu* perform ritual by applying turmeric and offering a chicken to the tree. After this ceremony only, the Geetari can extract toddy. As it is believed that the tree recognizes its legitimate owner, none other than the *Geetari* dares to climb the tree for toddy extraction with the fear that the tree goes dry or the climbers get serious injury. Plants and their parts play important magical properties and *Reddis* have been used as preventives in order to get rid of the diseases and misfortunes caused by the evil spirits (*Keedu*) in everyday life. Every house in the village can notice a bunch of dry red chilies and lemons (*Nimmanadantulu*) tied to the roof of the house at its entrance, which are believed to be chasing the intruding evil spirits into the house.

In the above, the cultural features of the sacred groves in the village of *Utalapalem* have been discussed, which are commonly owned resources. The *Reddis* use it and interact with it, for collective gain of the community. In this chapter, sacred groves are analyzed from the perspective of management of common group resources. The survival and utility of the common property is due to close dependence on natural resources. This makes survival more subject to a variety of unpredictable natural events that are likely to fall unequally in both time and space on the local population. If this inequality is threatening to a sufficiently large group, incentives may exit collectively managing it through exclusive use. The study village has number of commons used by the villagers for different purposes. They can be listed as follows: sacred grove, tamarind trees, mango trees, thati trees, canal, water stream, sand, forest, etc. The sacred groves are religious, common to the village in this region. These groves existing in the village provide psychological and sociological benefits to the village community.

The sacred groves of the study village are classified on the basis of productive and protective functions derived by the community. Hence, different type of groves existing in a village fulfils, the psychological as well as sociological requirement of the community. Therefore, sacred groves are incorporated in the local worldview. The village community has allocated an appropriate space to all the deities of their worldview as per their ascribed hierarchy and functions.

Benefits to the community from these groves are tangible as well as non-tangible. The village community looks towards the sacred grove deities for immediate as well as long-term support. The short-term benefits are in the form of support to production system. This is evident from the pattern of association community has developed through regular ritual process for all deities of the groves. Beginning with *Gangalamma* festival where all the village deities of different groves are offered obeisance, they at regular interval are worshipped in different festivals linked with economic activities of the community. This basically reflects the regular sustainable need-based approach of interaction with the deities of groves for surplus yield and protection of crops from external threat and pest problems. Therefore all those who are part of the village community, have developed faith in the groves and participate in its rituals and prescribed by the rules meant for *Gangalamma*.

The long-term benefits to the communities from sacred groves are in form of protection to human, livestock and production system of the village from the unforeseen incidents. It may be in the form if illness, natural calamities or epidemic. These incidents make the communities vulnerable and make them collectively solve the problems of distress. They seek solution from the deities of village living in these groves of the village. The mechanism used by them is by paying obeisance to the deities. For which they offer sacrifice to the deities of *Gangalamma, Bairudu, Parentalu, Chintalamma* etc at fixed interval, either once a year or after three or five years. This is basically for the long-term benefits incurred by the communities in the form of protection of human and livestock population of the village from diseases and external evil spirits. The description provided for mode of functioning of the deities living in the *Gangalamma, Bairudu, Parentalu,* groves, is that they constantly go round the village ridding on the back of a animal, keeping a watch and protecting the village from the attack of evil spirits and other calamities. This explanation shows the psychological support, communities of the village derive from the acts of deities of the groves. In return they protect the groves with a sentimental feeling of fear and respect. They have associated number of taboos, in order not to make unhappy the deities of the groves. These are intangibles benefits the communities derive from the groves.

Most of the studies treat the commons as a single use resource unit. But it is seen that the commons are put to multiple uses, like using it for fishing as well as bathing, communication etc. Common properties and sacred groves are commons put to multiple uses by the communities. Primary use of sacred groves is for conducting rituals spread through out the year to propitiate the village deities. But it has other uses like using it as common recreational ground as it is in the case of *Gangalamma Stalamu*. *Gangalamma* place that is located in the village settlements are used as playing, dancing and used for celebration of other ceremony. The *Goravadu* and other members of the village also collect produces from the trees of these groves as per prescribed norms. The medicine men of the village collect herbal species from these groves for treatment of diseases.

Analyzing sacred groves from the cultural perspectives shows that the community has internalized the groves in the cultural system. *Gangalamma* is incorporated in the religious system of the village. The *Soudi* is considered to be a place, where the village deities reside and provide various functions to the community. The God and Goddess of these groves are placed in the hierarchical order of the *Reddis* worldview and worshipped accordingly. The deities are collectively associated with protection and management of the village affairs. Different festival and rituals of controlling the unforeseen problems are assigned to them for the well being of the village. For these services they are propitiated at regular interval in the form of sacrifice, of poultry and animals in rituals and festivals. The *Reddis* have classified the deities residing in the groves on the gender lines. The deity of *Gangalamma, Nukalamma, Parentalu, Chintalamma* are female and deities of *Pothuraju, Bairudu, Rajulbabu* and *Adaviraju* are male. Though they are associated with different festivals and rituals, they collectively worship and offer sacrifice at all the festival and rituals associated with the common activities linked with village economy and well being of village people.

The economy of the *Reddis* is primarily agriculture based and rain fed. Activities of agriculture begin with the worship of *Gangalamma*. On the occasion of the *Gangalamma* festival all the village deities residing in different groves are offered sacrifice to be satisfied and perform their rituals. The agricultural activities proceed with the regular offering to the village deities at various intervals in the form of festival like *Bhumipanduga, Vanapandaga, Vithanalapandaga, Korrakotha, Papupandaga* and ends with *Varikothapandaga*. This linkage of village economy and religion through rituals in the sacred groves (common) for the well being of various economic activities are reflection of cultural incorporation of commons in the cultural system of the village. For instance, the *Korrakotha* festival is celebrated for offering the first produce to the village deities and then the households eat it.

All the villagers participate in sacred grove activities and festival celebrations. They distribute various roles in carrying out the festival activities at the groves. Participates in the ceremonies associated with the groves and make contribution of their share. The services are paid in kind for their activities. Therefore no activities either begin or end without worshipping the village deities. The villagers have constant fear of the spirit attack on the village therefore; they worship the deities of *Bairudu, Pothuraju, Adaviraju* and *Rajulbabu* for protection. They even worship and pray to these deities to protect them from natural causes.

In the case of sacred groves, the social solidarity and trust is exhibited by the collective celebration of festival and faith in the deities of the groves on different occasions. The *Reddis* abide by the taboos ascribed with the management of the groves. The solidarity is seen in participation of members in celebration of the festival. All the members attend the ceremony in the groves as per prescribed norms. The women and children are also involved in this process. The trust and reciprocity is seen in the firm decision of festival committee in the matter of contribution to be made for festival by every one. In this meeting the committee decides what should be done, how should be done and when should be done. Each household is assigned on rotation to contribute money so that it does not become a burden for few households. Similarly, every household contributes a small quantity of rice at the time of rituals, which is used in making the sanctified food and consumed by all the villagers. The unpredictable environment leads to development of trust among its members through religion to act collectively in order to cope up with the situation. The trust and solidarity exists in other affairs of the village, like the faith in the traditional council system of dispute settlement. The trust and solidarity is built through regular interaction in social affairs like participating in the household events of birth, marriage, death ceremonies and village festivals.

By the above discussion it shows that the sacred groves have passed through different phases and interpreting them from the dimension of successful management is not justified, as the macro level activities have influenced the outcome of the *Gangalamma* culture. But the resilience of the traditional communities shows their strength to protect their culture in changing context.

The anthropologist considered that all the elements of culture are embedded into one another. Sacred groves culture when seen from this viewpoint showed that the institutions of religion, economy, polity and social units are interlinked with the sacred groves culture. The grove is the entity and symbol of social cohesion for the community. The economic activities are undertaken after the collective choice agreement arrived at, after due consultation and during the celebration of different festivals in the groves. The culture of sacred grove has created a whole web of social network and system of management, which is interlinked with their social, economic, political and religious life of the *Reddis*.

For effective management of these groves, the community management approaches. Already the community is interlinked in the culture of sacred groves. Revival of the decaying institution will enhance the management of the village and itss resources. The recognition of the role of priest and other functionaries in the management of sacred groves only needs formal approval. Traditionally they are still functioning and are paid for their services by the community. The village also has formal institution Panchayat, functioning in the village. The Panchayat looks after the management of various development activities at the Panchayat level. The integration of traditional and formal institutions will lead to better management of the village social system and disbursement of development benefits to the community in the village. A sub-committee on sacred groves management should be created with the traditional functionaries giving them proper place to play the role in the management and conservation of the groves. By this formal recognition of the sacred groves, it will lead to better protection of the sacred grove areas and resources in it. Apart from this, the confusion in the community over the control of the sacred groves will be clarified, as it will get a formal recognition by the government through the local institution of governance. The integration of the two systems will provide a solution to present day conflict over the management of such resources in the village.

The tangible benefits are that they are common property, from where they collect various non-timber forest produce, herbs and shrubs needed by the households of the village and healers of the village. Some groves are used for dance, recreation purpose and conducting marriage ceremonies.

Common Property Resources and Management

Common property resources are sub-system of the larger social system and the participants of the common have multiple forms of involvement in a larger social group i.e. a wider community. The common property resources have certain features like property rights, ownership pattern, structure and function of institution for managing it and rules and roles for the communities follow for efficient use of the common resources. *Reddis* are fully conscious of their right over the common property resources, which are locally known as *Grama Kantam* in the village *Utalapalem*. The perception is that every household in the village has an equal and undisputed right over the common property resources. The villagers confirm that they have 12 tamarind trees, out of which 8 trees are within the village and 4 trees are outside the village boundary as common property resources. They have 150 Borassus palm, out of which 15 trees are within in the village and other trees are located outside the village as a common property resources. Villagers have 10 mango trees out of which 3 are within the village and the rest are under the reserved forest as common property resources. A canal (*Kaluva*), which runs through the fringe of the village, is a common property, the water of which is used for cultivation of paddy and other crops. A canal supplies water to nearly 35-40 acres of low land villagers and benefit 20 households of farmers. Since the canal water is equally distributed by each farmer, no conflict is noticed among the villagers. In the rainy season the water is also used for washing cloths. In summer sand (*Eshika*) available on the dry canal bed in the northern fringe of the village is used for house and check dam construction and for selling to other people of neighbouring villagers at a price. Other people buy a tractor load of sand for Rs. 20.00. Thus the money collected by selling sand goes to the common village fund. The stream (*Uvta*) water is used for meeting the needs of drinking water and washing cloths for the village people. The water is used sometimes for cultivation and for the purpose of bathing and drinking by the domestic animals. Forest, which is used for collection of minor forest produce, grazing domestic animals, and for other uses like fire wood collection, wood for house construction, agricultural implements, crafts, etc. are also common property resources. If the village people use these common properties there is no problem and there is no conflict among the villagers. But before using these common properties the villagers take the oral permission of the village headman or other people who are assigned the responsibility of managing these properties. If some body wants to construct house roof by Borassus palm tree leaves he has to ask the village headman and other people who manages. Generally village people do not ask any body if they want very less quantity.

Common properties like Tamarind and mango fruits are sold on auction in the village. Auction is locally called *Pata padadam*. The information is given to both villagers and the other villagers well in advance, usually one week before the date of auction is decided. People interested to take part in auction usually depend on the crop of the year. Both villagers and other villagers take part in the auction. The auction starts with a minimum price and gradually the price is increased until it reaches a reasonable amount. The money should be reasonable for both the villagers as well as for the auctioneer. The auction is declared final; the auctioneer has to pay the entire amount, which is agreed for at the first time of auction, no matter if he loses money eventually. The common properties also are sold among the villagers sometimes. The village elders may agree for a concessional price, if the seeker is from their own village. The amount is used for the celebration of village festivals and ceremonies. Sometimes this money is used for lending those who need to meet urgent needs, but the borrower has to return the amount with a minimum interest at the time of village festivals. If the borrower defaults, the villagers have the right to confiscate and auction the borrower's goats, oxen, cows or other properties to recover the borrowed amount. After auction of common property the auctioneer collects the fruits for selling either at one time or phase-wise.

To manage the common property resources under the supervision of the village headman the villagers select 2-3 persons every year among the villagers. Last year, the persons who managed the common property resources were Bachala Radha Krishna, Mamidi Satyanarayana Reddi, Rollupalli Bapiraju and Rollupalli Balaraju. In the year 2001, the villagers of Utalapalem received an amount of Rs. 1450/- from tamarind auction and the mangoes were auctioned at Rs. 850/-. In the year 2002 the villagers received Rs. 1000/- for the mangoes and Rs. 1000/- for tamarind. In case of any mismanagement by these selected persons, the villagers again sit together and select a new team to carry out the job. Villagers of Utalapalem rarely complain about the violation of norms set for using the common property resources. The villagers in a common meeting settle any conflict that comes up by chance. Warning is usually given to the violator, if any.

Money collected by selling the common property resources is deposited for the celebration of the village festivals like *Gangalamma panduga, Bairudu pandaga, Rajulbabu pandaga, Pappu pandaga, Kothamasa pandaga, Korrakotha pandaga*, etc. throughout the year. If the money is not sufficient for the celebrations, additional collection of Rs. 30-50/- per household is made. From this money they buy goat, pigs and hens for sacrificing on the festival days. They also buy coconuts, fruits, turmeric, *Kumkum* and other required materials for the purpose of performing rituals. For the festival of *Bairudu pandaga* every alternate year they sacrifice a goat and a pig. *Reddis* worship *Gangalamma, Pothuraju* and *Parentalu* for the peace and prosperity of the village and its people. They worship *Bairudu* for the prosperity of their domestic animals. They worship *Rajulbabu* and *Adaviraju* for peace between people and forest. The sacrificed animal is commonly cooked and eaten together by all. At the time of festivals they pay some money to *Goravdu* (*Pujari*) and *Bantrothu* (village messeger), and also to the drummer (*Baajawada*). Sometimes the villagers spend the required amount from this common fund to organize entertainment plays like *Gadasani Atta* for rainfall in the coming monsoon year. And the village council uses the money for other developmental works in the village like, road repairing, street lighting, etc. The villagers are alert and vigilant about the common property resources, as these resources generate money for them. Even free uses of these resources are done judiciously, so that availability of resources for everybody's need in the village is assured.

Reddis also had introduced new methods to collect money from the villagers, called *Kaida*, which is to be used for the developmental activities in the village. All the villagers at the time of selling their agricultural products, will have to pay some surcharge. The amount to be paid is 50 Ps per bag of brinjal; one rupee per bag of paddy, cotton, mangoes, tamarind, fibre (*Peesu*) and tubers (*Dumpa*); two rupees per bag of pulses; and five rupees per bag of cashew nuts.

Festival Associated with Livestock

Sankranthi is the festival celebrated for worshiping of the cattle owned by the villages. It is generally a remembrance of the services rendered by the cattle in agriculture production and by providing milk to the village households. It is held in the month of January. On this occasion, every household gives bath to their cattle by taking them to the water body in the village. Then they are decorated by anointing of horns with oil mixed in turmeric. After this they are taken to the grove and worshiped. The village deities are requested to keep them healthy, so that they keep providing the services to the human community. After the ritual is over, they are brought to the house and offered feeds, which is prepared for them.

Bairudu Pandaga is celebrated to propitiate the forest God (forest God), who resides on the Tenka Rai Chettu (*Ficus sp*) along with other plants. The forest deity of this grove is a male God called *Bairudu Devudu,* considered to be protector of the village domestic animals and people from sickness and other misfortunes from the forest. The main of this festival is to seek protection of the humans and the domestic animals from the wild animals. Poultry is offered every year, in all the festivals performed by the *Goravadu*. But the festival is celebrated once in 3 years on any Sunday in the month of August. Animal sacrifice is offered at regular interval for three years to appease the deity, He goat is offered to this deity in the name of celebrate *Bairudu Panduga*. The decision to sacrifice an animal is taken on the advise by the *Goravadu* depending on the intensity of the problem. They cook meals there, eat food, consume toddy (*Kallu*), and return to the village. This is a one-day festival and only men participate in the festival. The women and children are prohibited from entering the *Bairudu Panduga*.

Functionaries Dealing with Supernatural Powers

The head of household of each *Reddi* family or lineage is responsible for the propitiation of the spirits or deities of its family or lineage respectively. *Goravadu* is considered to be the priest and religious head of the village. He is considered the ultimate religious head of the village. The *Goravadu* has the right over the Gods and Goddess conducts rituals in festivals. In his absence some time the medicine man of the village can perform the rituals. He is believed to ensure cordial relations between the villagers and the supernatural powers.

Earlier the post of *Goravadu* was hereditary, but if due to any factor, the post of the *Goravadu* has fallen vacant, then it is filled through an elaborate ceremonial procedure. In this, every villager belonging to the community has right to get selected as *Goravadu*. The procedure has randomization element, where the deity herself selects a person as *Goravadu* of the village. *Goravadu* performs the sacrifices to propitiate the deities and spirits of the village who reside in the sacred groves. He is consulted in every religious matter and the villagers always take his guidance in different matters.

The *Vejju* is a black magician who plays the role of witch finder, witch doctor, spirit-finder and exorcizer. They know how to do it through observation though sometimes there are chance variations in the way of propitiation and prayer from one village or other.

Reddis believe in the propitiation and worship of a host of supernatural power related to sacred groves, which are supposed to be benevolent while others, those are malevolent and propitiated out of fear. The supernatural powers of sacred groves have local significance.

They also offer sacrifice of animals and fowls to the supernatural beings. They also make various magical observations during festivals and on occasions of risk or any sort of the rainwater from adjoining forest during rainy season flows into these lands, supernatural or natural danger that may threaten the individual, the family or the village community. The sacred groves, the place of *Devudu, Devatha, Parentalu* and the place of path and stones were some of the important sacred centers of the *Reddis* and are the places of practically all their religious activities. Besides these fixed centres there are also temporary representations of sacred centres having no fixed place but are located only at the time of need viz. during the performance of a *Gangalamma, Bairudu, Parentalu, Pendli,* sacrifice to ancestor spirits or such other religious ceremonies.

Village has prescribed pattern of payment in kind for the services he provides to village community. Every year, he gets four *Kunchalu* (one *Kuncham* is equivalent to four Kg) of paddy (*Vari*) at the time of harvest, from all the land owning households. A *Kuncham* is an iron vessel. Every household gives a fixed quantity of grain after the harvest. Besides this, *Goravadu* takes the entire offering made in ritual conducted in groves and other place, where he is asked to perform rituals.

The work of Bandroth acts as messenger to the village on the issues relating to village and sacred grove management. He also keeps watch on the common and private resources of the village, including groves so that no one causes damage or harm to them. If he comes across any such incident, should be reported to the village headman or priest of the village. They both try to take appropriate measure to punish the person or resolve the problem in village panchayat. Village headman generally belongs to the clan of first settlers in the village. He looks after the general management of the village and dispute settlement. Where as Goravadu also forms part of the village panchayat and his words have substantial weightage on the village issues.

The villagers cannot use the material available in the groves without permission of the priest. The priest has the right over the material of the groves that is dry wood fallen trees etc. it is believed that if anybody causes harm to the trees or things in the grove, the deity punishes him/her. This reflects in the form of sickness, crop damage to the perpetrator or dying livestock etc. immediately, he contacts the Goravadu, who suggested him with a solution, mostly it is performance of ritual in the grove, to conciliate the upset deity. Once the person performs the ritual, the deity forgives him. If some is seen causing harm to grove, the matter is reported to the priests, who impose fine on him for violation. Owing to fear of punishment by the deity, the person immediately pays the fine. In some cases, if the person does not pay the fine, the matter is brought to the village panchayat for settlement. The panchayat punishes the person with strict warning or fine. Priest cannot use the fine amount entirely for personal use but has to utilize a portion of it for the worship of deity at time of rituals.

If any tree falls down in the grove then the priest takes that tree. He can sell of the wood, and the amount obtained from it cannot be fully utilized by him, but the amount is used for village festivals. If the tree wood is used for construction or as fuel wood then priest has complete right over it. If some villager is in need of wood, then he can request the priest to give the wood to him, he can charges some money from him and gives the wood. The village members are afraid to interfere with groves of other villages and it is the case with their grove. This is because of the persistent fear of being harmed by the grove deity.

The monitoring of the grove is done in two ways. Firstly through the taboos associated with groves, where the influences of the deities are considered to check the misuse of the groves. Secondly, through functionaries like *Goravadu* and *Bandroth*, whose duties are to monitor these groves? The taboos associated with groves are mostly of prohibitive nature, and apply to people who are impure. For instance, women who are undergoing menstrual cycle are prohibited from going near to or inside the grove, and on all other occasion they are allowed to go inside the grove. Similarly those families, in whose house a death has taken place, are considered to be impure and prohibited from entering the grove. Since it is considered that impure person will pollute grove, which will annoy the deity of the grove and who in retaliation may harm the individual or village and its crops. Similarly, pregnant women avoid going near to or entering the grove, with a fear that deity of the grove may harm the child inside her. Along with this there is prohibition of extraction of dry woods and produces of trees in the grove, without the permission of the priest of the village. Generally the collection of herbs and edible roots, mushrooms, tubers are allowed.

In the case of other groves like *Bairudu* and *Rajulbabu*, taboos are of stricter nature and are based on sex and age. The women are generally prohibited to attain the celebration of festival. It is considered that groves may harm women and children. The monitoring by the village functionaries is done as discussed above. Here also, all the impurity related taboos are associated with these groves same as above.

The sacred groves are internalized in the local religious system, by arranging its deities in the hierarchy of the indigenous worldview. The village has its own sets of sacred groves associated with the villages economic, security and protection system. These deities residing in the groves are allocated separate common spaces in the village. Collective choice agreement of the village community is to regularly interact with them, through various festivals and rituals linked with various activities of the village community. This interaction, besides providing a common social occasion, also provides opportunity for information sharing and assessment of the status and condition of the groves.

Villagers depend on agriculture for livelihood. The collective provides them with information on activities of collective benefits like agriculture and security. This arrangement of regular interaction has created rules system, which specifies the time to begin and end the agriculture season, time for sowing, or harvesting, etc. For instance the collective interaction in festivals provide approximate information on availability of rain, and assurance through the appeasement of deities that they will take care of the impending problems. This gives the individual confidence that they can proceed with their regular activities, with out fear, as the deities of the groves will protect them.

Second, aspect of collective choice arrangement is associated with the issues of security to village system and protection to human life, crops and livestock etc. In the vulnerable environment were the impacts of natural calamities or diseases are interpreted in the form of behaviour of the evil spirit, and protection is sought from village deities against them, is an important psychological arrangement to cope up with uncertain environment at micro level.

The collective agreement of the village community is linked to the benefits derived by them from these groves, in an environment of unpredictability of nature. It provides them with security to their livelihood system, human and livestock. In reciprocation they regularly offer sacrifice to these deities of *Bairudu* and *Rajulbabu* groves. These ritual processes are collective affairs were all the male individuals of the village participate. Therefore when it comes to management of the groves a collective consensus is built within the village. The various festivals spread through out the year like *Gangalamma Panduga, Bairudu Panduga, Rajulbabu Panduga, Bhumi Panduga, Vithanala Panduga, Korra Panduga, Pappu Panduga,* etc, are collectively celebrated at village level, which reflects the collective choice arrangement of the local system, for the interaction with groves. At each festival, the villagers worship village deities residing in these groves.

This behaviour of the resource users is based on a sense of reasonably balanced sharing or reciprocity. Every user has relatively equal access to this resource, sometimes calculated according to means or needs, sharing it with his or her co-owners. Each participating individual is expected to provide assistance and support in managing the systems. In short, access to resources, and the social relation that grew up around them are developed through a collection of reciprocal rights, duties and privileges.

Settlement for dispute at intra village level, the village has a traditional council called panchayat. The aim of this Panchayat is to resolve the disputes taking place within the village system and maintains a harmonious living environment for the village. The nature of the disputes range from personal quarrels, family problems, marriage related disputes, property disputes, crop destruction by cattle, groves infringement, land encroachment, inter-community disputes, etc. At first the settlement of dispute is tried at family or neighbourhood level. The families try to settle the dispute within the lineage, but if unable to settle at these levels, then it is taken to traditional council i.e. village panchayat. Here the matter is heard at the village level and a collective judgement is delivered. If the conflicting parties do not agree to the judgement given by the panchayat, then they have highest dispute settling body in the form of Gram panchayat. The judgement of Gram panchayat is final and taken on the basis of unanimous consensus of the villagers and it becomes binding on the disputant to accept it or face expulsion from the community.

In Utlapalem village, the village council meeting is held at Ramalayam Temple. This place is located in the center of the village. Sometimes they used to conduct at community hall of the village, which is located beside the temple. The reason given for selecting a sacred site is that *Urupedda* is considered to be impartial in delivering the judgement, but if he is not, then the deity residing on the tree will punish him and bring disaster to him, family and property. It is essential that atleast one male member from every household is present in the village council proceedings, this is because the judgement is taken on consensus among the village members. It is ensured that representative from every household is present to give his opinion. Only male members of the village are allowed to participate in the village council activities. Women can only attend the meeting when they are disputant and have to put forth their viewpoint. But now women can also attend the meeting if they are interested.

Generally, an aggrieved person calls for the village council. He goes to every individual household and informs them about his problem and seeks their help in settling the dispute through the traditional village council. On the basis of severity of the problem villagers decides to call the village council. In general practice headman of the village belonging to the first settler family acts as *Peddakapu* of the traditional village council. The aggrieved party is called to explain his problem or incident and the person against whom the charges are made is also called to put forth his viewpoint. After hearing the viewpoint of both the parties, the council discusses on the issue, participated by all members present in the meeting. After hearing to all the members present of the village on the issue, the headman gives the judgement. Before delivering the judgement, the headman calls both the parties in the middle and then proclaims his judgement. The party, which found guilty, is given the punishment as per nature of crime, deeds and frequency of activities. The punishments are of two kinds, one is of positive nature i.e. reformist where the individual is asked to give communal feast, and other is negative i.e. penalty were punishment like monetary payment or social ostracizing is imposed on the person. In most of the cases monetary punishment is levied on the guilty person. The money is paid to the council, from which half of the money is given to the affected person to meet his loss. Rest half is deposited with the council for its usage. This happen in the case where both the parties accept the judgement, but in case either one of them do not agree to the council decision, then village council is called off by referring the matter to higher council called Gram Panchayat.

This region has a system of Gram Panchayat, which means that a conglomeration of eight to ten villages unit of association called Gram. Each Gram has a Panchayat, which settles the issues of dispute in its jurisdiction. Each Gram elects its Gram Panchayat through a democratically held election in which only the male members have right to vote. The term of a Gram Panchayat is for five years. This position is open to all the individuals within the Gram and no discrimination is made on the basis of tribe or caste for the candidate. Secondly a candidate can contest for the post as many number of times as he likes. Women are also allowed to contest or vote for the post of Gram Panchayat. Only men and women above sixteen are allowed to participate in this Panchayat. Members of other castes and communities are allowed to participate in the Panchayat.

The Gram Panchayat selects one representative in each village known as village member to assist him and advice him in the matter of dispute settlement. The village member is selected in consultation with the village members and in the absence of village headman the assistant adjudicates are the matters on his behalf. Mainly the assistant to Gram Panchayat is the *Urupedda* of the traditional village council. Gram Sadassulu and assistant village member, along with other village functionaries like *Goravadu* (priest), *Bandrothu* (village messenger) participate in the Gram Panchayat. All the members of the village under its jurisdiction attend Gram Panchayat. These days' women are also allowed to participate in the Gram Panchayat proceedings.

The disputant parties along with the headman of the village go to Gram Panchayat president's house and explain the total happening in the traditional village council Panchayat on the issue and request him to settle the matter. They fix time for the conduction of Panchayat. The meeting is held at the earlier place. On the fixed day, all assemble for settlement of the dispute. The Panchayat again hears the entire case on the issue and finally give his judgement. The judgement in this meeting is final and it is mandatory for the person or party to accept it or else he or she will be ostracized from the community. If any party disapproves the judgement by the Gram Panchayat, the fine amount is doubled, even then if the decision is challenged, then the person is ostracized from the community and the matter is closed. No such case has occurred in the villages till date and all the decisions have been accepted by the villagers and abided by.

In some cases, headman is not allowed to be the advisor of the Gram Panchayat. By doing so the villagers eliminate the possibility of his influencing the decision, where the advisor plays an important role in briefing the Gram Panchayat about the total incident. By keeping the option of random selection of judge, the villagers have provided every individual to seek fairest judgement he or she desires.

In the region six villages or eight hamlets come under this Gram Panchayat jurisdiction. Village *Utalapalem* come under *D. Ramavaram Grampanchayat* and *Yellavaram* constituency of East Godavari District. Gram Panchayat President is Rolupalli Srinivas who is 35 years old and lives in D.Ramavaram village. The village member of Gram Panchayat of Utlapalem village Pallala Vishwa Reddi 39 years old and lives in Utalapalem village. The tenure of village member is 5years. In this case the judgement given by assistant to Gram Panchayat is not mandatory for the disputant to accept it and they will not be penalized from being ostracized from the community if not agreed to it. Instead the matter is further referred to Gram Saba

Panchayat, who only has the right to exclude a person from the community.

In the process of judgement, generally it is not the conscience of the Gram Panchayat i.e. given importance but the collective will of the villagers present, who are considered as a member, who also participate in the discussion. In most of the cases the panchayat takes a unanimous decision on the issue and advices the Gram Panchayat. Based on the collective opinion of the Panchayat he announces the judgement. After this both parties are asked to put forth any grievances or request they have to make regarding the judegement, they can speak to the assembly. After hearing the grievances or the request amendment is made to the decision and proclaimed by the Gram Panchayat. This way the Gram Saba Panchayat functions as highest decision-making body in this region.

The taboos associated with grove are a mode of non-interference with the influence zone of the deities of the groves; which is correlated with their respective power and influence. But the villagers have designed certain flexibility in rules relating to use of these groves. These flexibilities are applicable only in a very extreme situation. The flexibility is linked with individual ignorance or village necessity.

According to the villagers of *Utlapalem*, if any one ignorantly breaks the rules of interaction with the groves, he has to only plead to the deity of the groves on his mistakes. But, if it is not an act ignorance then the deity punishes him/her. Than he has to seek the help of the *Goravadu* of the village and find the solution to protect himself from wrath of the deity. If any person in the village dire need of a product from grove like dried tree, or poles, then he approaches the priest of the village, who ask him perform rituals and make offering to the deity of the village and seek his permission. If the deity agrees, he is allowed to take the wood. This is judged in due course of time, that any untoward incident does not happens to the person, family or property, who had taken the wood. If any thing happens then it is interpreted that deity was against the extraction.

The sanctions are of internal and external type. The internal sanction relates to moral, guilt, curse, wrath of the deities, blessing etc., and external sanctions are punishment, fines, social outcasts, shames, etc. Both type of sanctions are used in the management of these groves. In the first case it is experimental. If an individual deviates from the norms prescribed for interaction with the grove, the deity residing in it punishes him, which is exhibited in features like sickness or loss of property or life, etc. in such a situation the individual seeks help of priest to know the cause and cure.

The level of sanction with respect to external sanction i.e. the rule associated with the management of the groves, is punished on the basis of degree of abuse of the groves and deviance from the rules. Generally in majority of the cases the villagers said that the deity of grove immediately punishes the person for his misconduct. Certain rules if abused, then the matter is taken to the village panchayat and the matter is settled by the traditional panchayat. The village takes the matter relating to the groves seriously and any person do not tolerate by it, can be eliminated from the community. Most of the time the punishment is of conformist nature, were the deviant is required to perform ritual in the grove and give a communal feast. The communal feasting is the process by which the individual makes it known to others that act he had done and to be pardon by them for the said variation.

Social Organization for Preservation or Conservation of Natural Resources

Other traditional taboos are scrupulously followed by the *Reddis* and this negative life sustaining restrictions facilitate production, protection and welfare of the community. They do not touch any crop or fruit either cultivated or available in the forest unless it is offered to their deities on the auspicious occasion. These celebrations are generally organized in such a way that all these crops, fruits, etc attain maturity and full growth.

As for preservation or conservation of the habitat and surrounding natural resources, Reddis strictly observe certain social norms. These norms regulate and guide natives in the process of exploitation of natural ecosystem and aim at conservation of natural ecosystem. This shows the natives awareness to the conservation and maintenance of natural ecosystem. The social norms emphasize the following view in the process of conservation of the natural resources. In the village every household is restricted not to over exploit the natural resources such as excess cutting of wood, edible leaves, fruits, tubers etc. According to them causing wastage, which may lead to scarcity of forest items. The villagers cut the timber only for their domestic purposes. They cut down forest wood to make agriculture implements such as plough, handle for certain other tools and for house construction. A part from these uses, Reddis do not specifically cut the wood for firewood rather, they collect the already slashed and half burnt wood from *Podu* fields. They only cut dry and fallen branches and bark of trees and utilize as firewood in their homestead. *Konda Reddis* believe that unnecessary and excess felling of trees and greenery lead to the destruction of natural resources, which ultimately may affect their livelihood. Another reason for not cutting wood for domestic purpose is that of the abundant availability of dry and fallen trees in the forest. Though Reddis cut the trees and forest growth on the hill slopes for *Podu* cultivation, they observe three years of fallen period for regeneration of forest growth. Reddis conserve the forest resources in order to practice *Podu* cultivation again in the same patch after three years. It is not that they conserve forest growth in the *Podu* to reoccupy and continue *Podu* cultivation in the same patch. Reddis possess indigenous knowledge about the utility of numerous plant species available in the forest. In addition to the conservation of forest growth in the forest and *Podu* fields, Reddis conserve most of the tubers and roots by leaving a small piece of tuber or root in the earth for further germination. So that to collect the same in the forth coming season. While collecting the broomsticks they select the fully-grown and matured plants and while cutting any tree they take care to see that ripened seeds fall on the ground for regeneration. At the time of bamboo cutting they cut one foot above or fully matured species and not tender ones. Similarly while picking the beedi leaves Reddis leave all the tender leaves to facilitate for growth of the plant into a big tree after which this tree yields fruits. A part from household level of conservation of natural resources especially land, forest growth and edible forest products, Reddis show concern about the conservation of fruit bearing trees and water resources at community level. In the study village the community owns certain number of jackfruit, tamarind, mango, cashew nut and caryota trees. The community as a whole takes the responsibility to safeguard them and takes measures to replace it with the same plant species in the case of damaged trees. Like this, Reddis as a whole attempt to conserve the natural resources.

It has been observed that *Reddis* are highly conscious about their local environment and resourceful plants and trees. In order to use in future, Reddis avoid uprooting of any tree from forest. They often cut the trees for any use but never uproot any tree unless it totally died. Generally they never destroy trees, which they use and even trees are protected in view of economic and ritual value in their subsistence pattern. To maintain the balance of ecosystem *Reddis* are involved in social forestry development by planting trees like cashew nut, palm trees, mango trees, tamarind trees, jack trees, jafra etc., where some trees have religious importance, while fruit bearing trees as well as tree which have economic value among them. They usually avoid cutting those trees in their niche.

Indigenous institutions among *Reddis* fall into the following categories: social, religious, political, judicial and economic. These institutions are organized on the basis of traditional roles and systems of authority, and legitimized in such structures as family, chieftaincy hierarchy, village council, and native or indigenous court systems. Appendix 20 provides key characteristics of indigenous institutions.

The study revealed several indigenous beliefs and practices that contribute to resource management. The earth and large bodies of water are regarded as sacred. A number of local taboos and prohibitions are related to practices believed to defile the earth. There are also rituals to seek permission from the spirit of the earth to dig or plough for the farm. There are also rituals to seek permission from the spirit of forest to cut the trees. The general belief is that failure to honour the spirits of the earth and forest or acknowledge its magnanimity in this way may provoke anger and vengeance in the form of natural calamities such as famine, excessive rainfall, or death of person. Appendix 21 provides information about indigenous beliefs and practices.

In this sense, indigenous institutions among *Reddis* have much to offer. Their philosophy of life is aimed at the perpetuation of all objects, both animate and inanimate. Their collective environmental wisdom and ethics are expressed through religious beliefs and a range of sacred and cultural practices. For example, the belief that the earth has a power of its own which is helpful if propitiated and harmful if neglected, is a powerful moral sanction against the wanton destruction of natural resources among *Reddis* region. The constant reminder of the good deeds of ancestors acts as a spur to good conduct on the part of the living. Similarly, the belief that lesser gods or spirits dwell in such natural resources as hills, forest, rocks, trees, canals and certain animals is tantamount to attaching intrinsic value to all objects-animate and inanimate.

7

Summary and Conclusion

From the preceeding discussions, it is understood that local knowledge is the knowledge that people have gained through from their ancestors. Over centuries, indigenous people around the world have developed their own, locality-specific knowledge and practice, which is an important part of the lives of the people. This knowledge has always been very important for the local people who generate it; it is a matter of survival for them. Indigenous knowledge is the local knowledge that is unique to a culture or society. It is also known as folk knowledge, people's knowledge, traditional wisdom or traditional science. Moreover, Indigenous knowledge is also a key element of the social capital of the poor and constitutes their main asset in their efforts to gain control of their own lives. Infact, indigenous people are the original inhabitants of a particular geographic location, who have a culture and belief system distinct from the international system of knowledge. Some scholars feel that such a definition is too narrow, in that it excludes peoples who may have lived in an area for a long period of time but are not the original inhabitants. This has led to widespread use of the term local knowledge, a broader concept that refers to the knowledge possessed by any group living off the land in a particular area for a long period of time.

It is also observed from various studies that over five decades of mainstream development, development strategies have tended to ignore the existing "knowledge-base" of the communities. Planners have designed development interventions that are entirely divorced from both the local reality and local expertise knowledge. In this manner, a vast resource of knowledge and information is rapidly getting lost, as local knowledge systems are primarily oral traditions. With rapid environmental, social, economic and political changes occurring in many areas inhabited by indigenous people, comes the danger that the IK they possess will be lost forever. Younger generations are acquiring different values and lifestyles as a result of exposure to global and national influences, and traditional communication networks are breaking down, meaning that elders are dying without passing their knowledge to children. In some cases, the actual existence of indigenous people themselves is threatened. As a result of this, there is a growing interest in the incorporation of local knowledge and traditions to provide environmentally sound approaches for development and people have started thinking the issue more seriously.

Subsequently, awareness of indigenous/local knowledge has been steadily gaining ground in the academic world, both within the social as well as in the natural sciences. However, interest in Indigenous knowledge was until recently confined to a few individuals and institutions. Dissatisfied with the outcome of previous developmental efforts, a growing number of scientists and development professionals are recognizing that indigenous knowledge provides a valuable resource for sustainable development. In due course of time, importance and contributions of indigenous knowledge systems and practices captured special attention from various range of disciplines such as biology, medicine (human and animal), pharmacology, agronomy, soil sciences and social scientists etc. In recent development theory and practice, many researchers all over the world are now striving to explore, know, record and use what the people know about using and conserving biological resources. Different studies also highlight the fact that local knowledge immensely contributes to the natural resource management systems, particularly in the developing world. In addition to this, recognizing, empowering and incorporating indigenous knowledge in participatory rural development projects has been considered a means of ensuring socially, environmentally and economically sustainable natural resource management.

Anthropologists and other social and natural scientists have written about the category 'indigenous knowledge' often make reference to the notion of 'value'. The approaches are in line with the concept of making local knowledge available and applicable to local self-reliance, decentralization of decision-making and fair access to natural resources. These involve the local population exclusively setting their own objectives, implementing the measures themselves on the basis of their own view and approaches to solving the problems and evaluating the results themselves. Utilizing IK in research projects and management plans gives it legitimacy and credibility in the eyes of both local people and outside scientists, increasing cultural pride and thus giving motivation to solve local problems with local ingenuity and resources.

Documenting the local knowledge systems of the rural population is one route towards recognizing the initiative of the community. This has been defined as a systematic body of knowledge acquired by local people through accumulation of experiences, informal experiments and intimate understanding of the environment in a given culture. It is unique to a particular group and local people such as farmers, who are custodians of certain kinds of knowledge. The potential use of it in agriculture is widely acknowledged, but lack of documentation on indigenous knowledge systems poses problems for researchers, policy makers and rural development practitioners.

Attempts have been constantly made in different areas to understand the patterns of indigenous knowledge/local knowledge and the domain of management of natural resources by the local people. Different researchers have tried to explain the conditions and the processes of collective choice, patterns of resource uses, and changes in it due to external influence in the course of history. Notwithstanding the conventional bias towards the superiority of the western science over the local indigenous knowledge, it is certain that the recognition of local knowledge gained significant ground when the application of scientific knowledge failed to bear the intended result due to social and cultural obstructions in the acceptance of rational knowledge, which sometimes stood in contradiction to the local knowledge. Despite the scholastic contribution of all these studies, it is still felt that the study of local knowledge system is by and large removed from the total cultural context in which it is located. Sometimes while local knowledge is construed as superstitious and an impediment to the process of development in less developed countries. Elsewhere, there is an uncritical clamour in favour of local knowledge system as the ultimate panacea for the survival of the forest communities. The majorities of tribes in India are indigenous and continues to practice and use indigenous knowledge in their day-to-day activities. The Indigenous Knowledge System National Programme of India is targeting this knowledge system, which has been largely neglected, relegated to the periphery and therefore remains largely unprotected. In general, India lacks an enabling and protective policy environment for this particular knowledge system.

To have a comprehensive understanding of the local resource management process, the present research emphasizes some aspects of "local environment knowledge and natural resource management" among the Konda Reddi of East Godavari district in Andhra Pradesh. The main objective of this study is to explore the Konda Reddi knowledge of agriculture, forest and forest produce, wildlife, etc. in their present form. Further, the study also throws light on the natural resource management of the Konda Reddi.

The present study adopted participant observation, direct observation, household schedule, in-depth interviews, informal discussions, focus group discussions, key informants and case studies, as research tools playing an important role in data collection. This helped the researcher to pick out and record issues. Initially, the study was to cover broad areas, which were derived from the review of the literature. These were knowledge of agricultural practices and natural resource management.

The study in the introduction firstly gave definitions and terminologies used in the context of knowledge of communities and conclude that the most generic term to be "local knowledge". Local knowledge derives from specific localities and refers to the day-to-day life of communities. The local knowledge system perspective when looked along with the management of natural resource perspective, offers well-sustained and embedded approach for resource management, which helped in sustenance of the society for long time through various cultural upheavals.

The sample village is located in Utalapalem village. It is inhabited by Konda Reddi. Government has recognized them as a Primitive Tribal Group (PTG) since the year 1980. Their basic livelihood is mainly depended on agriculture and forest resources. There are many excellent indigenous practices associated with their daily livelihood. For instance, the indigenous agriculture system had been based on the use of heap manure, mixed cropping and specific crop rotation. Organic manure enriched soil fertility while mixed cropping allowed a variety of microbes to flourish in the rhizosphere of different plant species, thereby enhancing the natural fertility of soil as well as helping pest control. Specific crop rotations were a boon to replenish soil nutrients. They possess elaborate knowledge on the type of land and quality of land needed for different types of crops to be grown. The farmers of this village categorize agriculture lands into four types, one is *Konda Podu* is the upland and dry land area where natural forest grows. The type of this land is suitable for the cultivation of local variety of paddy, maize, *Nuvulu*, small millets, pulses, chilly and vegetables. The second type of land is the dry land, which is called *Merakabhumi or garavubhumi*. It is low land, the rainwater from adjoining forest during rainy season flows into these lands. *Reddis* cultivate tubers, pulses, cotton, tobacco, vegetables etc. The third type of land is known as *Polam*, the rainwater from streams, adjoining forest during rainy season flows into these lands providing it with lots of nutrients and increases the fertility of the soil. These are suitable for only paddy cultivation, as it needs large quantity of water. They construct bunds (*Gattu*) for the purpose of water storage in these fields. The fourth type of land is called *Regadi* land in the vicinity of the village for growing jowar, pulses, vegetables and cotton. The varieties of crops grown by the villagers are described in the earlier chapter. The farmers of the village only cultivate single crop i.e. in Kharif season, and after that the fields are left fallow and it becomes a common grazing land for the cattle of the area.

Generally, farmers plan for the agriculture activities, with the help local knowledge, generated through decades of experiences, like quality of land, soil, rain, pest, etc. But some of them fall in the domain of uncertainty like nature of forthcoming rain and pest attack, because of which they plan for future ventures with a caution. In order to psychologically overcome the elements of uncertainty and unexplainable circumstances they seek the help of village deities and rituals. Agriculture in the region is totally rain fed. For the protection of crops, the village deities are propitiated to gain psychological confidence and satisfaction for the community to cope up with the uncertainty of nature. For this, they have designed rituals and ceremonies for various activities of agricultural management linked with different occasion of the agricultural activities. In course of these ceremonies, the villagers conduct some exercises to make metrological prediction like rain and future yield.

The *Reddis* farming method involve a large number of processes such as preparatory tillage, manuring, sowing, transplantation, weeding, reaping, threshing, winnowing, and finally transporting and storing up the products. The period involved in these processes varies from six to ten months according to the varieties of the crop.

Ploughing is a primary operation. In case of dry and wet land fields it is generally done by wooden plough yoked to a pair of bullocks. Spade, crowbar and digging stick are the important implements used for tillage of *Podu* fields and kitchen gardens. The number of ploughing varies with crops, for instance 3-5 times for paddy. All the farmers in the village are using wooden plough yoked to a pair of bullocks for field tillage.

Manuring in the field is done once in a year before the onset of the first monsoon in a year. The manure is spread over the soil, chemical fertilizers is either mixed with the soil before sowing or plantation, or spread over the crop when it is about one month old. Application of chemical fertilizers 2-3 times in the period between sowing or plantation, and harvesting. *Reddis* are yet to master the correct method and timing of application of different chemical fertilizers provided by the other people.

Seed sowing generally takes place after the field has been thoroughly prepared by ploughing and leveling. In case of paddy crop, the farmers have adopted the broadcasting method for sowing the seeds. In other crops like brinjal, cotton, pulses and vegetables is performed by dropping seeds in furrows made by spades horizontally along with long ridges that are made in the fields for the convenience of sowing as well as weeding. It is of interest to note that paddy plants that are grown in nursery beds are transplanted to the paddy fields when the seedlings are about ten to fifteen centimeters high.

Generally, weeding is done two to three times before harvesting. In paddy fields weeding is performed without any implements. In case of other crops it is done with either a hoe or digging stick. Harvesting generally takes place when the plants are fully grown in case of vegetables or crop is ripe in cereals, pulses etc. It is either done daily over a period as in the case of vegetables or carried out at the end of a period as in the case of paddy, pulses etc. Different types of implements are used for harvesting different crops. Whereas sickle is used for harvesting paddy crop and hand picking is done in case of vegetables.

Threshing and winnowing is necessary for paddy and pulses. First the paddy and pulses separated from the stalk with the help of wooden flail on a big bamboo mat or on the gunny bags. Winnowing of immature grain and other unwanted materials is carried out with the traditional implements like winnowing fan and winnowing spoon. After thrashing and winnowing they carry paddy to rice mill. If threshing and winnowing takes place in the field than the yield is usually transported by bullock cart, by men and women who participate in the operation.

Many wild plants, shrubs, weeds and grasses, are used as green manure in the field. Often, these are used as mulches and are later incorporated into the soil to improve productivity. Ashes of wood, husks and shrubs are spread on the soil in farms as manure. Animals such as pigs, rabbits, rats, squirrels and birds and insects cause major problems in the region. To control them, farmers use various techniques-traps, barriers and wrapping, hand-picking, spreading crushed leaves of some particular plants, burning, smoking etc.

Reddis maintain home and kitchen garden where various vegetables, fruits, flowers can be grown. Kitchen garden has become significant for self-consumption. Waste drainage water from the kitchen can be used to start seedling. Ashes and sweepings from the household and domestic animals' bedding are normally spread in the garden. The waste, water, ash, etc., and also the animal manure enriches the soil around the homestead.

Post-harvesting pest are not a problem among the *Reddis*, as most produce is consumed locally soon after harvesting. Small amounts of processed foods are stored by means of traditional indigenous techniques in special containers in an environment that is not easy for pests to attack. Farmers have to store their products, firstly for seeds, and secondly for consumption and sale. The most common method of storing paddy for consumption or sale is to have a separate storehouse for each family/household with a raised bamboo or wooden platform (*Gadde*). Farmers store unhusked paddy after threshing and sometimes apply some plant extracts as neem as insecticides. Storing paddy as seeds for the future is usually done in a different manner; the seeds are usually kept in a warm place in a container or in dry pumpkins.

Konda Reddis of this village have an old tradition of indigenous practices for maintaining their forests as a sustainable resource system. The forest management practices do not exist independently but are a part of the continuous flow of interdependent activities, which sustain Reddis in the village. The village forests are sacred and are controlled by the customary village council or chief who manage the exploitation of their numerous products. This reflects the high value given by local inhabitants to their forests. The village/ community forests in the region illustrate the way to manage natural resources through which we can understand the *Reddis* knowledge regarding management of environment. The important aspect of these activities is that they do not result in forest degradation, because collections and *Podu* cycle are controlled by the traditional management system. Among *Reddis*, it is traditional practice to preserve the forest in the immediate surroundings of the village and such forests are never cleared for cultivation purposes but are used as a convenient accessible source for the collection of materials for domestic requirements. Another practice of forest conservation is, making of certain areas as ritual or sacred forests (sacred groves) from where extraction of forest products is taboo. The whole community fallows such practices. Many traditions and customs are also given as the reason for protecting these sacred groves, and rituals. Sacrificial offerings are performed for the deities protecting these forests. Their beliefs and practices, related to sacred groves, have helped save many endangered plants in the forests. Another indigenous practice of the *Reddis* worth mentioning is the observation of good or bad days according to the moon for extraction of forest products like bamboo (*Veduru*), timber (*Teku*), palm leaves (*Tati akulu*), etc. It is believed that materials extracted on the bad days get attacked by insects and decay fast whereas those extracted on the good days do not. Such practice regulates the extraction of forest produce.

Bamboo plays an important role in the economy and is vital for supporting the livelihood of Reddis in the region. Bamboos are considered as multiple use, as they provide raw materials for food, medicines, construction of houses, farm implements, fence, baskets, weapons, utensils, fishing and hunting articles, fuel wood, handicrafts and so on. Tender leaves and shoots of bamboo are used as food and fodder.

Medicines made from herbs and shrubs have been helpful in the treatment of human ailments since time immemorial. Reddis of the village *Utalapalem* have rich tradition of medicinal use of plants; the knowledge has been passed from generation to generation through the oral traditions. A large number of plant species having medicinal values have been described in earlier chapter. The tradition of indigenous health care and medicine supported by a holistic concept of well-being is still practiced to some extent by the *Reddis* in the village.

Reddis identify an abundant variety of mushrooms. They are popular food, and many people are expert in the art of mushroom picking. According to them mushrooms are of varied shapes and size, colours and hues collected from the forest and some mushrooms have high medicinal value. Many species of such common edible mushrooms are collected from the forests of this region.

Wild plant resources provide for a variety of basic needs for the *Reddi* community. Building material, fuel, food supplement, materials for crafts, medicines are all acquired from the forests, and they also are a source of income. There are many varieties of non-timber or minor forest products, which are used for home consumption and are sold in the local markets. Selective collection of useful species of plants from forest for consumption and for a sale in the local market is one of the indigenous approaches. As the natural forest is an indispensable component in the region, both plain and hill *Reddis* are engaged in collecting, hunting, grazing and selective logging activities in the forests to supplement their needs and to earn cash income to supplement farm production. Plants collected from the forest include secondary staple foods, vegetables, fruits, flowers, seeds, tubers, mushrooms, bamboo shoots; in addition, many medicinal plants, fibres and weaving are also gathered from the natural forest. The *Reddis* of this village eat almost all wild animals, birds and some insects as dietary supplements.

Forest is a common property, so far as minor forest produce is concerned. Every individual of the village has equal right to collect MFP from the forest for selling in the nearest GCC. The proceeds from the MFP, although not impressive enough, help to supplement the family income; and more than that in ensuring the continuation of forest-based knowledge by the *Konda Reddis.*

Reddis classification of wildlife into animals, birds, reptiles, snakes, rodents, etc. are practically oriented. They know the sub-groups of each animal, their possible hideouts, general behaviour, reproductive cycle and the use of animal parts as medicine. They are aware of possible diseases that afflict the wild animals and the remedies for some of them. They can identify 27 varieties of snakes alone with their hideouts, characters (poisonous or non-poisonous), their eggs, and most interesting is their classification by the habit. Rats are harmful for *Reddis,* which prompts them to know details of rats and squirrels. Their knowledge of 61 varieties of birds is not less amazing. Most of these birds are eaten by *Reddis* Their knowledge of wild animals and their behaviour help them in devising appropriate hunting techniques. Annual hunting ritual for the *Reddis* is a colourful event with a ritual fervour. Its timing during the post-harvest summer months is most appropriate although they hunt any time they encounter a game. *Reddis'* association with the insects and flies is very intimate, as all of them share the common habitat. They classify insects on the basis of their mode of locomotion (flying, non-flying), colour, general characters, suitability of insects as food (edible and non-edible), and their utility (harmful and harmless), etc. They identify a long list of insects and flies, 23 of which are described. This list of 23 may be considered as a representative one, certainly not an exhaustive one. *Reddis* have the knowledge of their habitats, seasonal availability, reproductive cycle, nature and size of the larvae, etc. They are also very well informed about the bites of these insects and their cures. Although there is no tradition of worshipping insects and flies among the *Reddis*, they admit the importance of insects and their larvae as delicious edibles and as predators of many other harmful insects and flies.

In most cases, wild animals are classified and named at the species level. In addition, wildlife can be classified according to their value. The *Reddis* divide all wildlife into good ("themselves") and bad ("enemies"). The *Reddis* knowledge of the behavior of wildlife may be as detailed as the hunter/ gatherers. Their knowledge of wildlife behavior includes much attention to words the details of behavior and ecology of practically every animal. Animals and insects are used for food, tools and indicators of ecological dynamics. However, their consumption as food among *Reddis* is often accompanied by taboos that restrict the consumption or killing of certain species. For example, among the *Reddis* some animals are not eaten because of supernatural powers or because of ancestral taboos. Other wildlife rabbit, birds, insects, eggs of honey are eaten by all. The diversity in the use of animals is usually less than in plants. The knowledge of diseases and their cure is quite high among *Reddi* population, who have to depend on local means to mitigate the effects of diseases.

The management of natural resources in *Reddi* village is invariably integrated with local social institutions and religious beliefs to give them an ethical and spiritual look. Land, forest and water are symbolized as God and Goddess and worshipped by them as is trees, mountains, crops, and animals. The common properties, such as forest and forest produce, trees in the jurisdiction of village boundary and water are enjoyed with equal rights by every body in the village, which demonstrates the local democratic system. Community festivals, group hunting, and merry making symbolize a spirit of integration of the *Reddi* community with natural resources. Reverence for forest God as *Adaviraju*, for mountain as *Konda Devata*, and for water as *Gangalamma* is more from a religious and ethical code of conduct rather than from a mundane consideration of providing them with their lives' needs. This gives an impression that *Reddis*, given an opportunity that could be the best custodians of the natural resources, with which they are well versed and with whom they have a spiritual connection.

The sacred grove is also one such traditionally managed resource collectively owned by the community. The sacred grove perspectives and management of common property resources offer well-sustained and embedded institutions for resource management, which helps in the sustenance of the society for long time in history through various cultural upheavals. These are embedded in overall structure of the social system. The respective roles, rules, incentives and sanctions for management is gradually synthesized, evolved and acquired stability. The culture assigns, symbols sacred or profane as per the needs and utility of the social system to the natural ecosystem with which it interacts.

The village has sacred groves assigned to various village deities residing in them. They are *Gangalamma, Chintalamma, Bairudu, Pothuraju*, etc. The deity of *Burugu chettu* and *Vepa chettu* i.e. *Gangalamma*, is the principle village deity living in it. The deity living in *Tenka Rai* is *Pothuraju* and in *Juvvi cheetu* is *Bairudu* are male deity assigned the function of livestock protector from the forest. The deity of *Chinta chettu* i.e. *Chintalamma* is associated with marriage. The deity *Ganagalamma* is associated with the production system of the village i.e. agricultural activities and hunting. The protection of the village from natural calamities and spirits is also bestowed in the hands of village deity, for which they are propitiated at regular intervals of time. All the agriculture activities begin with worship in the grove, and end with a ritual and offering of the produces to the deity of the groves and then brought into use by the households. The village political system and other institutions are embedded into one another. This is evident from the interaction and role played by these functionaries in the management of the sacred groves and village affairs ranging from economic activities to protection of the people and resources of the village.

The villagers interact with these groves on the occasion of various rituals and festivals celebrated collectively at the village level. The villagers celebrate number of festivals in the village, namely *Gangalamma, Papu pandaga, Kotha masa pandaga, Korra kotha Pandaga, Bairudu pandaga, Pothuraju pandaga, Sankrathi, Dasara* etc. At each festival all the deities and spirits residing in these groves are worshipped and offered sacrifices of hen, cock and animals related to a particular deity or spirit. Only men folk are allowed to participate in the rituals conducted in the groves of *Bairudu pandaga, Pothuraju pandaga* and partake the sanctified food prepared in the groves. Community has evolved various taboos and prohibitions of different degree and kind in dealing with groves for men and women living in the village.

The sacred groves are internalized on the local religious system, and the sacred groves are associated with the village economic, security and protection system. Collective choice agreement of the village community is to regularly interact with them, through various festivals and rituals linked with various activities of the village community. This interaction, besides providing a common social occasion, also provides opportunity for information sharing and assessment of the status and condition of the groves.

Recapping the entire discussion shows that the study village has developed elaborate knowledge about the natural resources with which they interact, through observation in due course of history. The land for agriculture was classified as *Konda Podu, Merakabhumi or garavubhumi, Polam bhumi, Regadi bhumi* based on quality of land. They have their native ways of classification of plant and animal species. They use a number of plant species in their economic, healing and religious activities. In this context, the understanding of local practices of resource management and making use of it has become a need of the time. The sacred groves of the village are embedded in the culture of the village community. They govern and regulate the economic as well as the social system of the village. In the era of rapidly decreasing biodiversity and natural resources, such local practices of the communities need to be acknowledged and integrated into the management principles of the formal system. These groves are natural repository of local biodiversity as they have been conserved for a long time without disturbance. The linkage of agriculture and sacred groves is evident from the number of rituals associated with the different stages of agriculture work carried out by the villagers.

Reddis strongly believe that their culture and local knowledge are based on nature, and eventually, any loss of nature culminates in the loss of local environmental knowledge of the Reddis. The practical importance of local knowledge demands that the institutional setting where such knowledge is embedded needs to be strengthened, and people possessing such valuable knowledge need to be empowered.

The activities like watershed management should be encouraged in this region, as it lies in high altitude areas, where harvesting of ground water is not feasible. The region has facing water problem; therefore watershed management is one of the best options available for solving the water crises faced by the villagers for agriculture purpose.

The importance of local knowledge can be understood when one realizes that there are no rice or wheat plants nor cotton or mustard found lying around in the forest. What are found in the forest are wild plants and animals and moreover, communities has been depending on such resources over generations for their survival. These communities have bred out of the wild plants of the forests, the thousands of land races, which are the basis of the world's agriculture. The land races bred by farming communities are the foundation material of modern plant breeding and global food security.

In conclusion, ample evidence has shown that local knowledge provides a significant contribution to the survival of local communities, in terms of food production and security, crop yields, health care, and medicine. As the severity of environmental problems continue to increase, and solutions are sought to cope with such problems and improve the present environment, both scientific and local knowledge should be drawn upon to produce a good mix of policy options for local communities coping with some increasingly complex environmental problems.

Thus, the above information is highly valuable, as it has developed among the Konda Reddi community over generations. By understanding the local knowledge systems and communication processes of the people, we can begin to understand their needs and also their strengths. Therefore, it would be wise to protect and promote local environmental knowledge and wisdom widely for self sustainence in order to reduce poverty and hunger among indigenous people.

Appendices

Appendix 1
Shows Indigenous Agricultural Implements Used by Farmers

Local names	*English names*	*Operation*
Pedda Kathi	Billhook	Cutting bushes
Goddali	Axe	Cutting trees
Chinna kathi	Small knife	Cutting plants
Para	Spade	Collecting soil
Nagali	Plough	Tillage
Gadapara or gunapam	Crowbar	Digging holes
Butta	Bamboo basket	Caring goods
Gadde	Bamboo basket for storage	Storage grains
Kodavali	Sickle	Reaping
Gachella or voluva	Digging sticks	Weeding and sowing by digging sticks in *Podu* cultivation
Chata	Bamboo mat	Threshing
Balla	Wooden winnowing spoon	Winnowing
Edlabandi	Bullock cart	Transportation

Appendix 2
Provides Different Varieties of Crops Grown by the *Reddis* in their Fields

Local names of crops	*Local varieties in crops*	*Common name/scientific names of crops*	*Type of crops*	*Cultivation Land*
Konda Gantelu			Traditional	Podu
Konda Sollu or Sode	*Burre sode, chalaka sode*	*Eleusine coracana*/Finger millet	Traditional	Podu
Konda jonna	*Tella jonna or foreign jonna or gadajonna, pacha jonna or hybrid jonna*	*Sorghum Vulgare (Millet)/biclor*	Traditional	Podu
Konda korra or kooralu		*Setaria italica* /Italian millet	Traditional	Podu
Kond mokka jonna		Zeamaize/maize/Indiancorn	Traditional	Podu
KondaBontalu	*Golusu bontha and mamulu bontha*	*Echinocloa species*	Traditional	Podu
Konda Kandi	*Tella kandi, Yerra kandi, Nalla kandi*	Redgram or *Cajanus cajan*	Traditional	Podu
Konda sama		*Panicum miliare*/Little millet	Traditional	Podu
Arisanda/Alisanda			Traditional	Podu
Babbarlu	*Alasanda/u (low grain) arisanda/ alisanda, anumulu (low grain) chetubabara, chitti and mamulu babbara, police babbarlu*		Traditional	Podu
Arllu	*Neeti arlu, metta arlu*		Traditional	Podu
Minumulu	*Burada minumu, metta minumu*	Black gram	Traditional	Podu
Raagulu			Traditional	Podu
Aluvalu/ uluvalu	*Nalla uluvalu early crop/ mamulu uluvalu late crop*		Traditional	Podu

Contd...

Local names of crops	*Local varieties in crops*	*Common name/scientific names of crops*	*Type of crops*	*Cultivation Land*
Peasra	*Pilli peasra, metta pesara, konda pesara not cultivating*	Green gram	Traditional	Podu
Nuvvu /Nuvulu	*Pyrnuvi late crop in December, setakottu nuvi, tolakarinuvvi early crop in June, all Nuvulu (jonna fire), mamulu Nuvulu*	Ceasamum	Traditional	Podu
Vari	*Swarna masoori, samba masoori, chaitanya, karnataka swarna, punaasa, esselwa, khobari puvu, pearl or PL, khichidi cahitanya, bontha khichidi, Rella khichidi, Gidda masoori, punasa, vaazram, Krishna katuka, yakaraatri punasi, budumalu (metta dyanum), neetikasara, 64, 072, 105,1001, 1003,1010.*	Paddy or *Oryza sativum*	Traditional and modern	Traditional variety in podu Modern variety in modern
Pathi	*Mcus, tulasi, Jk2, bharani.*	Cotton	Modern	Plain land
Vanga	*Gulabi, Mulla Vanga, Tella Vanga, sarala, kranthulu, neetivankaya in this two nalapu and telapu.*	Brinjal or *Solanam malongena*	Modern	Plain
Dumpalu	*Karrapendalum, Faram, Kerala, Tella Gulabi (chilakadi dumpa) Eraa Gulabi (pedda pendalum dumpa), chama dumpa, mantum dumpa, shenda dumpa, kanda dumpa.*	Tubers	Modern	Plain

Contd...

Local names of crops	*Local varieties in crops*	*Common name/scientific names of crops*	*Type of crops*	*Cultivation Land*
Gidi mamidi		Cashew nut	Modern	Podu and plain
Pogaaku	*White barely, Kalipogaaku in this two one is Potharam and Garapa*	Tobacco	Modern	Plain
Konda Goggu		*Cochlospermum sp.* / Fiber	Traditional	Podu and
plain				
Gomu		Fiber	Traditional	Podu
Pandiri pandlu/ tamatalu		Lycopersicon esculentam		
Gongura		*Hibiscus sabdaiffa/* Rosselle	Traditional	Podu and
plain				
		plant		
Benda		*Hibiscus esculentus*/Okra	Modern	Plain/podu
Adusulu		Guizotia species	Traditional	Podu
Thiveralu		Kind of pulses	Traditional	Podu
Konda Budama		Traditional paddy	Traditional	Podu
Kulamalu		*Cucurbit sp.*	Traditional	Podu
Dhamalu		*Mucuna sp.*	Traditional	Podu
Dippalu		*Lagenaria sp.*/Bottle gourd	Traditional	Podu/Plain
Dhoki		*Lagenaria sp.*/Bottle gourd	Traditional	Podu/plain
Gummadi		*Pumpkin*	Traditional	Podu

Appendix 3
Provides an Overview of Various Local Rice Varieties Cultivated by the *Reddis*

Local rice variety name	*Local name land and soil type*	*Spikelet characteristic*	*Maturation time and types of locale June-December*
Swarna masoori, people are cultivating more good quality	*Polam* (wet land) and soil is light black and pasty (*giguru*) height of the spikelet is two and half feet	Seed is short and stumpy, spike let is short and not much strong and it is light yellow colour	Five months crop and gives 35 tones production
Samba masoori this also cultivating more very good quality	Same as above	Seed is small and thin and the spikelet is like *swarna* but white in colour	Five months and production is 25-30 tones
Chaitanya cultivating now also	Same as above	Spikelet and seed look like pearl variety and stem is stronger than *swarna* and three feet long	Six months crop and production is 22-29 tones
Karnataka swarna this aslo cultivating	Same as above	Seed is medium size and the spikelet strong in yellow colour with 3 feet long	It is four and half month crop and the production is 25-30 tones
64 this variety is cultivation	Same as above	Seed is biger than samba and spikelet is same in white colour with two and half feet long	Four months crop and production is 25-30 tones
072 also cultivaing	Same as above	Seed is long and stumpy pure white colour and the spikelet is *chaitanyna* type with three feet long	Five months crop and the production is 20-25 tones

Contd...

Local rice variety name	*Local name land and soil type*	*Spikelet characteristic*	*Maturation time and types of locale June-December*
1010 it is also cultivating	Same as above	Same as above	Four and half months crop and the production is 30-35 tones
Punasa it is old variety 20 years back they used to cultivate	Same as above	Seed is long and thin and the spikelet is *swarna* type in yellow colour and height is three and half long	Six months crop and the production is 10-15 tones
Esselwa is also old same as above variety	Same as above	Same as above	Six months crop and production is 10-15
Kobbari puvvu old same as above variety	Same as above	Same as above	Six months crop and production is 10-15 tones
Pearl or PL they cultivating now	Same as above	Seed is long and thin and spikelet is long and strong with white colour 3 feet long	Six months crop and production is 22-30 tones
Kichidi chaitanya cultivating	Same as above	Same as above	Six months crop and production is 20-27 tones
Bontha kichidi old variety	Same as above	Same as above three and half feet long	Six months crop and production is 10-18 tones
Rella kichidi old variety	Same as above	Same as above	Six months crop and production is 10-15 tones
Gidda masoori is cultivating	Same as above	Same as samba and spikelet is light white colour with two and half feet long	Five months crop and production is 22-27 tones

Contd...

Local rice variety name	*Local name land and soil type*	*Spikelet characteristic*	*Maturation time and types of locale June-December*
Yekaratri punasi old variety twenty years back	Same as above	Same size seed like samba and spikelet come out in one day only and it is three and half feet long	Six months crop and production is 10-15 tones
105 it is pld variety but now also cultivating	Same as above	Same as above	Five months crop and production is 30-35 tones
Vazram old variety	Same as above	Seed is round shape and the spikelet is like samba in pure white colour three and half long	Five months crop and production is 20-25 tones
Punasi old variety not cultivating	Same as above	Same as above	Six months crop and the production is 10-15 tones
Krishna katuka old variety 10-15 years back	Same as above	Seed is bigger than *swarna* but in coners of the seed it is black in colour and spikelet is 3 feet long in light black colour	Six months crop and production is 20-25 tones
Budumalu it old variety 30 years back some of them are still cultivating	Broadcasting	Seed is big in size and spikelet is *swarna* size red in colour	4 and half months crop and production is 8-12 tones
Neeti kasara old variety 20 years back not cultivating	Broadcasting any place	Swampy seed big size and the spike id very long 4 and half feet long light green in colour	4 and half months crop and production is 8-12 tones

Contd...

Local rice variety name	*Local name land and soil type*	*Spikelet characteristic*	*Maturation time and types of locale June-December*
Nalla konamanu old variety same as above	Same as above	Same as above	Six months crop and production is 8-12 tones in up lands
Bobbuli gante old variety like above	Same as above	Same as above	Six months crop and production is 8-12 tones
Vankasonnalu old variety same as above	Same as above	Same as above	Same as above
Veduru aku sannam old variety same as above	Same as above	Same as above	Same as above
Metta masoori old variety but still using	Same as above	Same as above	Four months crop and production is 20-25 tones on up lands
1001 now cultivating	Same as above	Big size seed and the spike let is like *karnataka swarna* 3 feet long pure white in colour	4 and half months crops and production is 25-30 tones
1003 same as above	Same as above	Seed is bigger than 1001variety and spike let is same as above	4 and half months crop and the production is 25-30 tones

Appendix 4
Provides Vegetables and Fruits Grown by the *Reddis* in their Kitchen Garden

Local names	*Local varieties*	*English name*
Mirapa	Konda Mirapa, Seema Mirapa in this two yerupu and telupu, kodi vesala	Chili or *Capsicum anum/ solanum sp*
Tamata	Hybrid Tomato, mamulu Tomato	Tomato or *Lycopercicum esculentum*
Beera Kaya	Pedda beerakaya, cheturu, hybrid.	Ridge gourd
Benda Kaya	Yeduaakula (seven leaves), Pedda Benda (ten leaves)	Ladyfinger/*hibiscus esculentus*
Chekudu	Gumpu Chekudu, kanasuposala, potti, maddi, gori, Chetti, Anapa, sriram, kurra bobbari, golgonda.	Beans or *Dolicus lab lab*
Anapa Kaya	Peduga, dippa (potti).	Bottle gourd
Donda Kaya		*Coccinia sp.*
Potla Kaya		Snake guard
Gummadi Kaya	Burada gummadi, mamulu gummadi	Pumpkin
Kakarakaya		Bitter gourd
Munnakaya		Drum sticks
Dosa Kaya	Kura Dosa, pandu Dosa.	
Mamidi	Varieties given in previous report.	Mango or Mangifera indicia
Eedae pallu	Variety of orange	Citrus sp
Panasa Kaya Jack fruit		*Artocarpus integrifolius/*
Parinda Kaya, papaya, bobas Kaya		*Carica papaya*/Papaya
Nara mango		Wild mango
Seeta palam		Custard apple
Ram palam heart		*Anona reticulata*/Bullock
Saema		*Colacasia sp/colocasia*
Nimmakaya		Lemon
Chinta Kaya		Tamarind

Appendix 5

Provides Konda Reddi Division of Labour in Agricultural Activities

Operation of work	*Gender*
Selection of *Podu* fields	Adult male
Slash chap/cutting trees and bushes	Adult male
Clear bushes and stones	Adult male
Leveling	Adults
Covering seeds	Adult women
Burn bushes	Adult male
Ploughing Podu land	Adult male
Digging holes	Adults male and women
Broadcasting	Adult male
Sowing	Adult women
Paddy transplantation	Adult female
Weeding	Adult women
Guarding crops	Adults male and aged men
Construction of shed for protection of crop	Adult male
Harvesting	Adults male and women
Preparation of basket for storing grains	Adult male
Gathering harvested crops	Adults male and women
Cutting corn	Adults male and women
Paddy Thrashing	Adult male
Preparation of thrashing ground	Adult male
Paddy Winnowing	Adult male
Measuring	Adults
Preserving	Adults male and female
Tuber cutting	Adult women
Fencing	Adult men and women

Appendix 6
Shows Reddis Division of Labour Gender wise

Male–Tasks	*Female–Tasks*
Selection of *Podu*, clear *Podu* fields: slash, chop, burn, cultivation of *Podu*: digging holes, broadcast seeds, guard the crop and harvest	Clear and collect small bushes together plant, weed winnow and harvest.
Settled cultivation of fields: ploughing seedling, weed, guard the crop and harvest.	Thrash the ploughed field, cover the seeds with sand, transplant, weed, winnow and harvest
Foraging: collect wild tubers, roots and mushrooms, collecting honey	Collect wild green leaves, tubers, roots and mushrooms, collect firewood.
Hunting: hunt, trap, skin	
Miscellaneous: sell minor forest produce, build hut and prepare fence, fashion wood and prepare iron implements and prepare brew caryota toddy	Sell minor forest produce pound grains, fetch water and cook, prepare fence.

Appendix 7
Provides Different Varieties of Edible Green Leaves Identified and Consumed by the Reddis

Local name	*Scientific name*	*Available months*
Bachala Kura		June-May
Boddi Kura	*Cissamples pariera*	August-October
Botha boragala Kura		August-October
Chama Kura		August-October
Chary Kura		June-May
Chenchali Kura	*Derris scandens*	July-October
Dola kondlem Kura	*Mucuna pruriens*	June-May
Goddu Kura *or* bhramatantem	*Bauhinia sp.*	March-May
Gurugu Kura/Guruvakura	*Digeria muricata*	June-May
Kokkeli Kura,	*Clorophytum tuberosum*	June-May.
Konda thota Kura		June-May
Kondlem Kura		June-May
Mandam Kura	*Hispida sp*	July-October
Pitta kanda Kura	*Psychotria monticola*	July-October
Pullagumadi Kura	*Corniculata sp.*	June-May
Seekum Kura	*Acacta sinuate*	June-December
Thota Kura	*Amranthus viridis*	March-May
Ubasaru Kura/Ubbakura	*Cardiospermum halicacabum*	March-May
Vaska Kura		December-May
Goguchettu akku pachadi	*Cochlospermum sp.* Pickle	March-May

Appendix 8
Provides Different Varieties of Edible Mushrooms Identified by the Reddis

Edible mushrooms	*Location of sprout*	*Available months*	*Consume in the form of*
Adda kokkulu	sprout on rotten adda tree	July-August	Curry & soup
Amidakokkulu	sprout on castor decaying	July-August	Curry & soup
Annekokkulu	sprout on rotten wood	July-August	Curry & soup
Bedi-Bedikokkulu	sprout on rocks	July-August	Curry
Budide kokkulu	sprout on ash soil	July-August	Curry & soup
Chirmanu kokkulu	sprout on chirumanu decaying	July-August	Curry & soup
Chiru putta kokkulu	sprout on anthills	July-August	Roasted
Gadda kokkulu	sprout in the fields	July-August	Roasted
Gaddi kokkulu	sprout on harvested paddy decaying	July-August	Roasted & Curry
Garuva kokkulu	sprout on dead trees	July-August	Curry
Godugu kokkulu	sprout on plain ground	July-August	Curry
Jonna kokkulu	sprout on maize decaying	July-August	Curry
Mamidi kokkulu	sprout on mamidi decaying	July-August	Curry & soup
Moddukokkulu	sprout on rotten wood	July-August	Curry & soup
Mollam kokkulu,	sprout on sandy soils	July-August	Curry & soup
Nelakokkulu	Sprout on sandy soils	July-August	Curry & soup
Pandi kokkulu	sprout on Manure and damp places	July-August	Curry & soup
Pedda putta kokkulu	sprout on below big anthills	July-August	Curry & soup
Pidisikokkulu	sprout on holes of the trees	July-August	Curry & soup
Sakaku kokkulu	sprout on plain ground	July-August	Curry & soup
Veduru kokkulu	sprout on bamboo decaying	July-August	Curry & soup
Pidem kokkulu: Pidem	sprout on bamboo groves	July-Sept	Curry & soup

Appendix 9
Provides Different Varieties of Non-edible Mushrooms Identified by the Reddis

Non-edible mushrooms	*Location of sprout*	*Available months*
Chinta kokkulu	sprout on tamarind tree decaying	July-August
Tati kokkulu	sprout at Thati tree	July-August
Panasa kokkulu	sprout on the braches of jack tree	July-August
Etha kokkulu	sprout at Etha tree	July-August
Mollam kokkulu	sprout at Mollam tree	July-August
Maddi kokkulu	sprout on Maddi tree decay	July-August
Gunde kokkulu	sprout on plain areas	July-August
Pachoru kokkulu	sprout on Pachoru decay	July-August
Burugu kokkulu	sprout on Burugu tree decay	July-August
Oddise kokkulu	sprout at oddise tree	July-August
Tanggudu kokkului	sprout on Tanggudu tree decay	July-August
Nemali kokkulu	sprout on Nemali tree decay	July-August
Vyapa kokkulu	sprout on Vyapa tree decay	July-August
Bodda kokkulu	sprout at Bodda tree	July-August
Dunna kokkulu	sprout on Dunna tree decay	July-August
Mudda kokkulu	sprout on Manure	July-August
Verri kokkulu	sprout on paddy field	July-August
Rai kokkulu	sprout at Rai tree	July-August
Saama kokkulu	sprout on Sama decay matter	July-August
Kovelli kokkulu	sprout at kovelli tree	July-August
Gupen kokkulu	sprout on huge decay wood	July-September

Appendix 10
Provides Different Varieties of Edible Tubers Identified and Used by Reddis

Local name of the plants	*Plant scientific name*	*Available months*
Aarethega	*Dioscorea oppositifolia L. Dioscoreaceae*	April-May
Adavikanda dumpa	*Dioscorea sp.*	January-June
Adavitega dumpa	*Dioscorea sp.*	November-January
Bajinika/Banika	(Vanda tessellata) Roxb G.Don (Orchidaceae)	January-Febuary
Chauti dumpa		June-May
Cheda dump:		June-May
Donda dumpa	*(Solena heterophylla Lour.) Cucurbitaceae*	January-June
Geetum dumpa		January-May
Gintega	(Dioscorea glabra Roxb. (Dioscoreaceae)	Febuary-March
Nalla Gadda	(Diocorea bulbifera L. (Dioscoreaceae)	October-December
Nara dumpa	*(Dioscorea wallichi Hook) Dioscoreaceae*	March-May
Nela Gadda/ Kaluva Dumpa	*Nelumbo nucifera Gaertn* (Nelumbonaceae)	April-May
Nula Dumpa	(Dioscorea tomentosa Koenig ex Spreng. (Dioscoreae):	April-May
Pindi dumpa or *pidi dumpa*		November-January
Putta Kanda	(Amorphallus paeonilifolius) Dennst. (Nicoloson (Araceae):	Febuary-March
Racha vymu dumpa	*Pimpinella heyneana.*	June-July
Salva Dumpa	(Lasia spinosa L.) Thw. (Araceae)	March-May
Sare dumpa	*Desmodium sp*	June-May
Savida dumpa	*(Dioscorea puber* Blume) Dioscoreaceae	October-December
Tamara Gadda	(Smilax zeylanica L.) Smilaceae	January-Febuary
Teega dumpa	*(Naravelia zeylanica L.)*	June-May
Thegadumpa	DC. (Ranunuculaceae)	
Thene Guridi	(curculigo trichocarpa) (wight) Bennet & Raizada (hypoxidaceae)	
Vayamu dumpa	*(Dioscorea pentaphylla L.)*	March-May
Velichanda Gadda	*(Dioscorea hispida Dennst. (Dioscoreaceae)*	October-December

Appendix 11
Provides Various Types Fruits, Seeds, and Leaves Collected by the Reddis

Local name of the plants	*Plant scientific name*	*Plant Part used*
Addachettu	*Phanera vahli*	Fruit seeds are edible found in summer
Allichettu	*Memecylon edule*	Fruits are edible found in summer, colour of fruit is red.
Baranika	*Streblus asper lour*	Fruits are edible
Baranika	*Streblus asper lour*	Fruits are edible, Latex used for curdling milk, stem twigs used as tooth cleaning.
Bodda chettu	*Ficus hispida*	Fruits are edible found in summer
Cheepuru pandlu	Grewia sp.	Fruits are edible found in summer
Chintakayalu	*Tamarindus indica*	Fruits are edible found in October-May
Chitra chettu	Solanum sp.	Fruits are found in summer, unripe fruits used in preparing chutney.
Garaga Modigachettu	*Butfa monospenna*	Fruits are edible found in summer
Girumanu chettu		Fruits are edible found in summer
Glass Manu or *Juvvi*	*Ficus arnottiana*	Tree gives white colour flower found in summer consumed by the birds
Gotti chettu	Zizyphus xylopyrus	Fruits are edible found in summer
Jaaruga chettu	*Buchanania lanzam*	Fruits are edible found in summer
Jambo neradi	*Myrtus cyneinum*	Fruits are edible comes in rainy season
Jarugu chettu		Fruits are edible found in summer
Jeedi koora	Clausena heptaphylla	Leaves used in culinary as substitute for curry leaves.
Kinneri chettu		Wild animals like *Duppi* (dear), *Kanasu*, and *Gorre* eat the bark of this tree.
Konda karivepaku	*Murraya sp*	Tender leaves used for preparing chutney and seasoning curries.

Contd...

Local name of the plants	*Plant scientific name*	*Plant Part used*
Konda mamidi	*Mangifera sp.*	Raw fruits are used as vegetable.
Konda tangedu/ Adda pikka	*Xylia xylocarpa*	Seeds roasted and eaten
Kovvelchettu	*Sterculia urens* Roxb.	Roasted seeds are consumed.
Kusuma/Boosi	*Schleichera oleosa*	Fruits are edible
Maredi chettu/ maredu	*Aeglemermetos*	Fruits are edible found in summer.
Nallachedi	*Semecarpus anacardium*	Fruits are edible found in winter season
Neredichettu	*Syzygium cumnini*	Fruits are edible found in summer season
Pala chettu	*Manilkara heyandra*	Edible fruits found in February
Panasachettu	*Artocarpus interrifolia*	Fruits are edible found in summer
Pardonda chettu		Edible fruits for birds found in summer
Parmi chettu	*Zyziphus oenoplia*	Fruits are edible found in winter season
Peepara kayalu	*Cucumis sp.*	Fruits are edible found in winter consumed by cattle.
Pitta pisirika	*Flacourtia sp.*	Fruits are edible
Reguchettu	*Mauritiana lamk*	Fruits are edible found in summer season
Seemaneradi or *Kondaneradi*	*Syzygium sp.*	Fruits are edible available in rainy season
Tenka Rai		Fruits are edible found in summer consumed by Birds
Thanguduchettu	*Cassia auriculata*	Fruit seeds are edible found in summer.
Usiri	*Phyllanthus emblica*	Fruits are edible found in summer
Velaka or *Velaga*	*Limonia clephantum*	Fruits are edible found in winter season.

Appendix 12
Provides Various Types of Grasses and Plant Products Used as Fodder

Local /Scientific name	*Area of location*	*Grass size*	*Available months*	*Part used for*
Yerra shangala	Found in black soil, waste lands, fields corners (Gattulu)	1 foot long red stem	June-Febuary	Grass used as fodder
Pala shangala	Dry land same as above	White stem	June-Febuary	Grass used as fodder
Boddu thonga or Pedda thonga (*Cyperus rotundus*)	Wet lands	3 foot long and spread out 10 branches	Throughout the year	Do not east because of tasteless
Tella tumma (*Acacia lucophloea*)	Dry lands (Meraka bhumi)	3-10 foot long	Throughout the year	Leaves used as fodder
Nalla tumma (*Acacia nilotica*)	Dry lands	3-10 foot long	Throughout the year	Leaves used as fodder
Jeebu thunga (*Spinifex littoreus*)	Dry lands	1 foot long green stem	Throughout the year	Grass used as fodder
Nara tunga	Wet lands	1 foot long thin	June-January	Grass is tasty used as fodder
Garki gaddi (*Cynodon dactylon*)	Dry lands	1 foot long green stem	Throughout the year	Mostly ate by horse and donkey
Avisa (*Sesbanum grandiflora*)	Dry lands	3-10 foot long	Throughout the year	Leaves used as fodder
Adavi tumma (*Canavalis gladiata*)	Dry lands	3-10 foot long	Throughout the year	Leaves used as fodder
Kanpa gaddi	Wastelands (Metta prantham), ditches (Garuvu), banks of water bodies.	1 meter long spread 8-10	Throughout the year	Grass tastes good used as fodder

Contd...

Local /Scientific name	*Area of location*	*Grass size*	*Available months*	*Part used for*
Neeti arika	Wet lands, Paddy fields (polam)	1 foot long spread 4-8 branches	June-February	Grass used as fodder
Yenugu Mokalu (*Elepahantopus scaber*)	Wet lands, Paddy fields (polam), banks of canals	1 meter long green stem spread 4 branches	Throughtout the year	Grass used as fodder
Aggiantapu kada	Wet lands, Paddy fields (polam), banks of canals	1 meter long with single branches	Throughtout the year.	Don't eat due to bad smell
Ippa chettu (*Madhuca indica*)	Dry lands	3-15 foot long	Throughout the year	Leaves used as fodder
China garica (*Eragrostis tenella*)	Dry lands	3-15 foot long	Throughout the year	Leaves as used fodder
Iddi gaddi (*Heteropogan contortus*)	Dry lands	1 foot long	Throughout the year	Grass used as fodder
Madunu Kada (*Borreria hispida*)	Dry lands (metta prantham)	1 foot long red stem spread 8-10 branches	Throughout the year	Plant used as fodder
Velaka chevi Kada (*Ipomoea eriocarpa*)	All kind of places	1foot long spread 6 branches	June-February	Plant used as fodder
Garika gaddi (*Cynodan dactylan*)	Dry places	I foot long	Throughout the year	Grass used as fodder
Jerri gaddi	Dry places	1 foot long	Throughout the year	Very tasty grass
Pyru gaddi (*Arthraxon sps*)	Dry lands (metta prantham)	1 foot long	June-January	Grass used as fodder

Contd...

Local /Scientific name	*Area of location*	*Grass size*	*Available months*	*Part used for*
Chepala Gaddi (*Polyalthia cerasoides)*	Dry lands (metta prantham)	1 foot long	June-January	Grass used as fodder
Gokkeri Gaddi	On bunds (gattulu) of paddy fields	1 foot long	June-August	Grass used as fodder
Nulaka Gaddi (*Helicteres isora)*	Dry lands (metta prantham)	$1^{1/2}$ meter long spread 40-50 branches	July-November	Grass used as fodder
Dabba Gaddi (*Cymbopogon citrates)*	Sandy soils, black sticky soils (regadi)	1 meter long	June-December	Grass used as fodder
Erruthota gaddi (*Ischaemum indicum)*	Dry lands	1 foot long	Throughtout the year	Grass used as a fodder
Rella Gaddi (*Cassia fistula)*	Sandy soils and wastelands	2-10 meters long	Throughtout the year	Tender leaves used as fodder
Chepuri Gaddi (*Thysanolena maxima)*	Bedi (rock bed) and in the fields	1meter long spread 4-30 branches	Throughtout the year	Tender leaves used as fodder
Lady Gaddi as fodder	Wet lands	1 foot long	Sept-December	Tender leaves used
Bodha Gaddi (*Cymbopogan coloratus)*	Dry lands in the fields	1 foot long	Throughtout the year	Tender leaves used as fodder
Nimma Gaddi (*Cymbopogan citrates)*	Wet lands in the feilds	1 foot long	Throughtout the year	Tender leaves used as fodder

Appendix 13
Provides List of Some Endangered NTFPs Identified by the Reddis

Local name	*English name*	*Local name*	*English name*
Vacha	*Acorus calamus*	Guggilum	*Boswellia ovalifoliolata*
Maredu	*Aegle marmelos*	Saliva dumpa	*Lasia spinosa*
Naramamidi	*Litsea glutinosa*	Errachitramoolam	*Plumbago inidica*
Errachandanam	*Pterocarpus santalinus*	sarpagandhi	*Rauvolfia serpentine*
Srigandham	*Santalum album*	Tella karaka	*Terminalia pallida*
Erragadda	*Ceropegia*	Radam	*Dendrobium herbaceum*
Galli chettu	*Entada pursaetha*	Miriyalu	*Piper nigrum*
Pedda julapala	*Heterostemma*	Nalla tegada	*Operculina turpethum*

Appendix 14
Provides Information About the Plants Used by the *Reddis* for Construction, Furniture, Implements and Fibre

Local and scientific name of the plants	*Uses of plant products*
Somichettu (*Soymida febrifuga*), Addateega (*Bauhinia vahlii*), Alli Nemaliare (*Vitex altissima*), *Annem chettu*, Chelaka dideri manu (*Polyalthia cerasoides*), Panderga chettu, Lolika or ollika chettu (*Stereospermum xylocarpum*), Neredu (*Syzygium cumini*), Yegesa (*Pterocarpus marsupiumaa*)	Tree wood or log is used for the construction of house erect pillars
Chenum chettu, Dudiippachettu (*Hymenodictyon excelsum*), Gesari, Gummudu (*Cisscus pallida*), Ippachettu (Madhuca longifolia), Maddi (*Morinda bracteata*), *Glorigoja chettu*, Cherumanu (*Anogeissus latifolia*), *Yelma*, Thella Karra (*Terminalia chebula*), Kukka Bodda, Bandare or Daddugu manu (*Adina corniculatum*), Rellu Gaddi (*Saccharam spontaium*)	Wood is used for construction of house roof.
Nallamaddi (*Terminalia alata*), Nallamaddicheetu (*Terminalia alata*), Neradu/Nerodi chettu (*Syzygium cuminu*)	Tree wood or log is used for the construction of house attic"
Oddise (*Cleistanthus collinus*), *Olica*, Pachori (*Dalbergia paniculata*), Regu/Renu (*Mauritiana lamk, Shannenu*)	Wood is very strong and used for cross beams in house construction and leafy branches are used for fencing."
Tanguduchettu (*Cassia ariculata*), Tuminika (*Ospyrus melaoxylon*), Nallamaddi (*Terminalia alata*), *Clenam*, Vepa (*Azadirachta indica*), Neradi (*Syzygium cuminu*)	Tree wood is used for the construction of house side/walls"
Karaka chettu (*Torminalia chobola*) Yetukarra chettu, Errdditta chettu, Panasachettu (*Artocarpus interrifolia*), Veduru (*Bambusa aroundinacea*), Dirisana (*Albizzia lebekka*), Kagu (*Dalbergia indica*), Guggilum (*Shorea robusta*)	Used for house construction.

Contd...

Local and scientific name of the plants	*Uses of plant products*
Aanem chettu	Tree wood used for construction of first house pillars
Teku (*Tectonia grandis)*	Wood is very strong and use for house construction
Thatiakku (*Borassus palm)*	Tree leaves are used for thatching houses.
Dakkagaddi (Cymbopogon citratus)	Grass used for thatching houses of roof
Veduru pidum (*Dendrocalmus strictus)*	Sticks are used for both roof, side walls and fencing of the house
Thogerika chettu (Morinda pubescens, Aanuda chettu	Fencing
Nallagantha (Diospyros sylvatica)	Wood is used for the construction of cattle sheds.
Addaaku (*Bauhinia vahlii)*	Leaves used as meal plats. Stem bark used as fibre.
Cheepuru pandlu (*Grewia sp.)*	Stem bark used as fibre for making ropes.
*Boddachettu (*Torminalia chobola)	Tree jute is used as fibre for making rope
Goguchettu (*Cochlospermum sp.*)	Extract jute for making rope,
Tati Balegelu (Borassus flabellifer)	Used as fibre
Jeelugu naara (*Caryota urens)*	Fibre is used for making bow and arrow.
Guddelem Nara and *Kitta Nara*	Fibre used to tie the tobacco leaves to hang for drying the leaves
Nulaka Gạddi	Grass used to make ropes to tie bullock cot
Adda nara (*Bauhinia vahlii)*	Tree fiber is used for construction and also manufacturing rope
Tati Nara (*Borassus palm)*	Leaf petiole used as a fiber for construction
Pemu	Tree fiber is used for construction
Juvvi (*Ficus arnottiana)*	Tree fiber is used for construction

Contd...

Local and scientific name of the plants	*Uses of plant products*
Thada chettu *(Helicteres sora)*, *Chakka manu*	Wood is used for manufacturing bed pillars.
Teku (*Tectonia grandis)*	Wood is very strong and use for furniture and house construction
Lolika or ollika chettu *(Stereospermum personatum)*	Wood is used for preparing bed.
Bandaru or Daddugu manu (*Adina cordifolia*), Girumudi, Adavi teku (*Tectonica grandis*) Bandaru (Adina cordifolia), Gummudu (*Gmolina arborea*), Doodugu	Wood is used for making house doors and for other furniture's.
Gummadi teku *(Gmelina arborea)*, Tada *(Grewia tiliaflora)*, Pachori *(Dalbergia paniculata)*, Gannera *(Samanea saman)*, Ganneri Manu *(Samanea saman)*, Errodi manu (Rosewood), *Gesari*, Vepa *(Azadirachta indica)*	Tree wood is used for manufacturing cot and furniture
Theddu pala	Tree wood used for furniture and utensil like spoons
Burugu *(Bobaz ceiba)*, Chindugu *(Albizia procera)*	Wood is used for manufacturing match sticks
Plants are used for manufacturing implements and Tools	
Neradi/nerodi *(Syzygium cumunini)*, Rella chettu (*Cassia fistula*) Yegise *(Pterocarpus marsupium)*, Bella *(Abutilon hirtum)*, Mokken, Rella *(Cassia fistula)*, Vepa *(Azadirachta indica)*, Korimi (*Ixora arborea)*, Vandanam, Sunera *(Ochna obtuse)*, Tangedu *(Cassia auriculata)*, Gummadi *(Gmelina arborea)*, Marri *(Ficus bengahalensis)*, Nalla tumma (*Acacia nilotica*), Chinta (*Tamarindus indica*), chiru manu (*Anogeissus latifolia*), Gumpena (*Lannea coramandelica*), Sandra (*Acacia catechu*)	Tree wood is used for in manufacturing plough

Contd...

Local and scientific name of the plants	*Uses of plant products*
Veduru *(Bambusa aroundinacea)*	Used for knife, sickle, axe, spade hoe etc.
Gummadi *(Gmelina arborea)*, Chindugu *(Albizia procera)*, Vandanam karra, Vegara, Nalla maddi *(Terminalia alata)*, Bandaru *(Adina cordifolia)*, Vegisa *(Pterooarpus bilous)*, Billa *(Chloroxylon swictenia)*, *Andugu* *(Boswellia serrata)*, Gesari, Vandanam	Wood is used manufacture bullock cart and its parts
Peddapala chettu *(Tribulus terrestris)*	Wood is used for grinding implements
Tellamaddi chettu *(A.leucophloca)*	Wood is used for rollu
PeddaPalachettu *(Manilkara heyandra)*, Bella Manu *(Abutilon hirtum)*, Korimi chettu (*Ixora arborea)*	Wood is used for crusher
Bamboo (*Bambusa aroundinacea)*	Stick of bamboo is used for bow and arrow
Jeelugu marra *(Caryota urens)*	Fibre is used for string of bow
Jeelugu naara *(Caryota urens)*	Fibre is used for tying string of bow.
Bamboo (*Bambusa aroundinacea)*Juvvi (*Bambusa aroundinacea)*	Used for making baskets
Balli chettu *(Blepharis madaraspatana)*	
Pedda gummudu (*Gmolina arboroa*)	Wood is used for making musical instrument like *Thontam, Gilaka,* guitar Veena etc.
Ganneru Chettu, Adaviteku (*Tectonica grandis*), Errodimanu (Rose wood), Vepachettu (Neem)	Wood of these trees is used for manufacturing tables, stools, and chairs.
Gesarichettu	Timber is used to make small wooden stools known as peeta, which is used for stting purpose.

Appendix 15
Provides Information About Forest Plants Used as Fuel

Yelma or velma chettu (*Limonia clephantum*),
Palachettu (*Manilkara heyandra*),
Tangedu chettu (*Cassia auriculata*),
Somi chettu (*Soymida febrifuga*),
Maddi chettu (*Mentaptera arjuna*),
Oddisa chettu,
Pilli manu (*Maerua oblongifolia*),
Vepa chettu (*Azadirachta indica*),
Usiri chettu (*Phyllanthus emblic*),
Chitipalu chettu,
Kakitechettu (*Argyreiasetosa*),
Vegisachettu (*Pterooarpusbilous*),
Korimi (*Ixora arborea*),
Pardonda (*Glycosmis pentaphylla*),
Karaka (*Terminalia chebula*),
Panta,
Palloriteega,
Edurutanda,
Bella (*Abutilon hirtum*),
Neradi (*Syzygium cumunini*),
Ippa (*Madhuca longifolia*),
Gumpen (*Lannea coromandelica*),
Mulla (*Barleria prionitis*),
Rella (*Cassia fistula*),
Chinta pala (*Tamarindus indica*),
Kampugumudu,
Chirumanu (*Anogeissus latifolia*)
Oddise,
Neradu (*Syzygium cumunini*),
Veduru (*Bambusa aroundinacea*),
Kinneri chettu,
Gumpenachettu (*Lannea coromandelica*),
Thogerikachettu (*Ceriops decandra*),
Korimipala chettu (*Ixora arborea*),
Palachettu (*Manilkara heyandra*),
Pardonda chettu (Glycosmis pentaphylla),
Nallagantha chettu (*Diospyros sylvatica*),
Glass Manu or Juvvi (*Ficus arnottiana*),
Tatta chekum
Nalla tumma (*Acacia nilotica*)
Dirisana (*Albizzia lebekka*)
Simachinta (*Pithecolobium duice*)
Barniki (*Streblus species*)

Appendix 16
Provides Information about Diseases Treated by Means of Medicinal Plants Collected from the Forest by the Reddis

Name of disease	*Name of the tree, herb and Useful part*	*Processing*
Errakamerllu (Red jaundice)	Errapatti mokka (*Mimosa pudica* root) and 3 Miriyalu (pepper)	Made paste of ingredients used to cure fever.
Tellakamerlu (White jaundice)	Tella Patti mokka (*Mimosa spp* root) and 3 Miriyalu (pepper)	Made paste of ingredients used to cure fever.
Kadapunoppi (Stomach pain)	Sonti little salt and chilly powder	Made Powder used as medicine
Kadapunoppi (Stomach pain)	Bark of *pillichettu* 3 Miriyalu (Pepper) Mix with Bellam (Jaggery)	Made paste of ingredients used to cure pain
Gontupattukovadam (Throat infection)	Sonti with *manchi tulasi* herb	Made paste of ingredients used to cure infection
Kamerllu (jaundice)	Root of Ducharichettu (*Achyranthes aspera*), ballatantemchettut, kisikindagaddi, kolliveru chettu, seethamma chettu, kummari chettu, mixed with *pachi passapu* (turmeric), *Ravadichettu bark* and *Miriyalu* (Pepper).	Made paste of ingredients used to cure fever
Jwaramu (Fever)	Root of yeduakulapala chettu (*Alstonia sctolaris*), *pathalageradi chettu*, and *Konda kasiminda* mixed with *pachi passapu* (turmeric)	Made paste of ingredients used to cure fever
Jwaramu (Fever)	Root of Vempali chettu (*Tephrosia Purpurea*) and 3 *Miriyalu* (Pepper)	Ground made paste used to cure fever
Jwaramu (Fever)	Bark of dakki chettu (*Oroxylum indicum*) mix *pachapasapu* (turmeric), 3 *miriyalu* (Pepper)	Make paste used along with water

Contd...

Name of disease	*Name of the tree, herb and Useful part*	*Processing*
Daggu (Cough)	Fruit of *mamulu karka* tree, munnuru karkaya tree (*Terminalia chebula*)	Made paste prepare tablet used as medicine
Daggu and *Jwaramu* (Cold and fever) among kids	Fruit of *gara chettu*, *mamulu karka* tree, munnuru karkaya tree (*Terminalia chebula*)	Made powder used with mother milk
Headache	Root of *sukandi teega*	Ground and used as medicine
Cold among elders	Root of mullukaya (*Solanum surattense*)	Mixed in curry used as medicine
Verochanalu (motions among kids)	Milk of mamulupalachettu (*Holarrhena antidysenterica*)	Extract milk and used as medicine
Verochanalu (motions among elders)	Root of Janteneru teega (*Clitoria ternatea*)	Made paste and used as medicine
Verochanalu (motions among elders)	Leaves of nelavemu (Andrographis paniculata	Ground made tablet used as medicine
Verochanalu (Loose motions)	Bark of peddamanu (*Ailanthus excelsa*), *nemalimanu*, mix with Turmeric	Make paste used as medicine
Jeernamkaniappudu (Upset of stomach, or any indigestion problem)	Bark of yetimaddi/tellamaddi/peddamaddi (*Terminalia arjuna*)	Wet in water and drink
Kadapu and chati noppi (Stomach and chest pains) among elders	Nalla jilleduchettu *(Calotropis gigantean)* tellajilleduchettu (*Calotropis procera*)	Extract milk and apply on infected parts
Aayasam (asthma)	Root of palloriteega (*Tribulus terrestris*), and *shannengi* tree bark (*Lagerstroemia paryiflora*), *pachipasapu* (turmeric)	Ground and made tablets used as medicine

Contd...

Name of disease	*Name of the tree, herb and Useful part*	*Processing*
Jomma disease (fits).	Bark of *nalleru chettu* and *badademanu*, 3 *miriyalu* (Pepper) and *pachapasapu* (turmeric)	Make paste and used as medicine
Ubbu and *noppi* (Swell and pain)	Fruit sap Nallagidi/ nallachidi (*Semecarpus anacardium*)	Extract sap and apply on infected part
Knife cuts, Burns, wounds	Extract Leaves juice Aggikura (*Tridax procumbens*)	Extract juice used to cure infected wounds and cuts
Ubbadam (Healing wounds)	Root nelathadi (*Curculigo orchioides*) add little salt	Ground and used to cure infected wounds
Kathigiyadam (Cutting wounds)	Extract sap from leaves bokkena chettu (*Phyla nodiflora*)	Extract sap used to cure cuts
Keedlanoppi (Joint pains)	Extract bark milk nidaraganneruchettu (*Vinca rosca*)	Extract milk apply on infected parts
Keedlanoppi (Joint pains)	*Aarechettu* fruit seeds	Ground and apply on pains
Saygadda (Heat boils)	Bark of munakkaya/munaga (*Moringa oleifera*)	Heat the bark and apply on boils
Gaddalu (Normal boils)	Leaves juice velturuchettu/bedturu (*Dichrostachys cineria* mix turmeric powder	Extract leaves juice apply on infected part
Kalakadlu (conjunctivitis)	Extract milk from the petiole of the leaves of Pichikusumu/ballirekkachettu (*Argemone mexicana*). And swallow the fruit seeds of *Gulimindachettu*.	Extract milk and apply in the eyes
Rakthasula (blood comes from tooth)	Bark of yegesichettu/ vegisachettu (*Pterocarpus marsupium*) mix with drops of hen blood	Ground and mix drops of hen blood apply on infected part
Stomach infected by small worms	Bark of Vyapachettu/ vepachettu (*Azadirachta indica*)	Ground and take with water to cure

Contd...

Name of disease	*Name of the tree, herb and Useful part*	*Processing*
Kukkakaravadam (Dog bite)	Root of *tella jilledu* chettu mix with *nuvvulu nunne* (sweet oil)	Ground and apply on infected part
Pamukuttadam (Snake bite)	Extract juice from leaves Nallaisari mokka (*Aristolochia indica*) mix with *miriyalu* (Pepper) bark of Erra guliminda chettu (*Abrus precatorius*)	Extract juice take with water to cure
Daiyam and pedda pichasi (Evil Spirits)	Bark of *peddamanu, kakkite, sirthalabudi, ratigante* mix with *Elluli* (Garlic)	Make powder and give with water
Pundlu (boils)	Oomatha *(Datura sp).*	Extract juice from leaves and apply on boils
Kuthikelu (Migraine and mumps)	Gurrapu Dekka *(Mucuna monosperma)*	Make paste of seed used as medicine
Gayalu (Internal wounds)	Pitta pisirika (*Flacourtia sp)*	Stem bark pounded and juice orally administered for one teaspoonful thrice a day.
Abortion (*kadapu thisukovadam*)	Guruvinda (*Abrus precatorius)*	Made Seed paste used orally
Centipede bite (*jerri pothu*) given orally for insect bite.	Raavadi chettu (*Dillenia sp.*)	Stem bark pounded with turmeric and
Wound cuts (*katti debbalu*) and dysentery (*Virochanalu)*	Paala *(Holarrhena antidysenterica)*	Latex used for cuts and wounds, stem bark used grounded and taken for dysentery one spoonful for three times.
Tamerlu (skin diseases)	Rella chettu (*Cassia fistula)*	Fruits are grounded & used to cure disease.
Skin disease called *Gajji*	Bogade chettu	Bark grounded and used as medicine.
*Adavi Gali (*forest evil eye)	KampuGummudu (*Ampelocissus tomentosa*)	Bark grounded & used orally as medicine.

Contd...

Name of disease	*Name of the tree, herb and Useful part*	*Processing*
Chest pain	Rella chettu (*Cassia fistula*)	Stem bark grounded with jaggery and taken orally.
Pannu noppi (tooth infection)	Ippa chettu *(Maduca longifolia)*	Calyx edible, seed oil used in culinary, latex used for tooth infection, stem bark juice dropped into the left year if tooth decay is on right side.
Kadapu manta (ulcers and gastric) and skin diseases	Kusuma/Boosi *(Schleichera oleosa)*	Fruits grounded used for ulcers & gastric trouble. Seed oil used for skin diseases.
Snakebite (pamu kaatu)	Pamu chettu *(Pergularia sp)*	Root grounded and used as medicine.
Snake bite	Paathala garida *(Rauvolfia serpentina)*	Fruit grounded and used as medicine.
Stomach disorders (*Kadapu noppi)*	Pandu pilli *(Diospyros sp)*	Stem bark grounded and made juice used for.
Speech impairment (*Nathi*)	Musidi (*Strychnos nux-vomica*)	Stem bark ground with turmeric and one teaspoonful administered for patient.
Swelling (*ubbu*)	Nallachedi *(Semecarpus anacardium)*	Seed juice is applied as medicine.
Menstrual disorders	Peda thanthemu and Paampana (*Oroxylum indicum*)	Stem bark grounded and orally administered to the patient.
Heat *(Vedi)*	Sugandhipala *(Hemedesmus indicus)*	Root juice used for cooling effect.
Fracture joints *(bokkalu athukovadam)*	Naramamidi *(Polyathia tonffolia)*	Bark grounded used as medicine
Blisters and skin disease and itchiness. (*Gajji Tamara*)	Chedukurra *(Andrographis sp)*	Leaf paste applied to cure.
Cold (rompa)	Seepachettu *(Crinum defixum)*	Leaf juice used as medicine.
Axe, Knife, and Sickle injuries	Galidonka	Leaves of this plant are crushed in the palm & the juice is applied on the wounds.

Appendix 17

Discusses the Economic Importance of Various Wild Plants and Products Identified and Used by the *Reddis*

Local name of the tree	*English Name*	*Parts collected*	*Sold@per Sack/Kg*
Guggelem Chettu	*Aegiceras corniculatum var floridum*	Black gum	Rupees 60/- Kg at G.C.C.
Kovvel Chettu	*Sterculia urens*	Gum	In this there are three types pure white gum Rs 140/- Kg light white gum 100/- Kg and black colour 60/- Kg. at G.C.C.
Gumpena chettu	*Lannea coromandelica*	Gum	Rs 20-25/-Kg at G.C.C.
Konda gogu	*Cochlospermum sp.*	Gum and Fibre	Fibre is Rs 5/-katta and gum is 60-130/- Rs. Kg at G.C.C.
Somi chettu	Swietenia febrifuga	Gum and fruits	Fruits Rs 8/- Kg and gum is Rs. 50/- Kg at G.C.C.
Puthika chepurlu	*Dphniphyllum glaucescens*	Broom Sticks	One broom Rs 4-6/-at G.C.C.
Konda chepurlu	*Grewia salvifolia*	Broom Sticks	One broom Rs 6-12/- at G.C.C.
Usiri Chettu	*Ambla or Emblica officinalis*	Fruits	Rs 20/- kg at G.C.C. One bag Rs. 200/- at private business man
Indupu Chettu	*Strychnos potatorum*	Fruit and seeds	Rupees 3-4/- Kg at G.C.C
Musidi Chettu	*Strychnos nux-vomica*	Fruit seeds	Rs 2.50 Kg at G.C.C
Karak Kaya Chettu	*Terminalia chebula*	Fruits available	Rupees 5/- Kg at G.C.C. medicinal value
Chinta Chettu	Tamarind (*Tamarindus indica*)	Fruits and fruits seeds	Fruits with seeds are Rs 5-10/-Kg and without seeds are Rs 10-20/-Kg; fruit seeds are Rs. 3-6/-kg at G.C.C.
Nalla chidi	*Semecarpus anacardium*	Fruit seeds	Rs. 3-4/-kg at G.C.C.
Ippa chettu	*Madhuca longifolia var latifolia*	Fruit seeds	Rs 4/- Kg at G.C.C.

Contd...

Local name of the tree	*English Name*	*Parts collected*	*Sold@per Sack/Kg*
Konda thadi	*Terminalia bellerica*	Fruits	Rupees 4/- Kg at G.C.C.
Nunegeche	*Caesalpinia bonduc*	Fruit seeds	Rupees 3/- Kg at G.C.C.
Noone ghakha or *latifolia*		Fruit seeds	Rupees 5/- Kg at G.C.C.
Chikkum chettu	Acacia caesea	Fruits	Rupees 5/- Kg at G.C.C
Maredi chettu	*Aegle marmelos*	Fruits	One bag Rs 100-200/- at G.C.C.
Naramamidi Chettu	Wild mango (*Polyalthia longifolia*)	Bark	Rs 12/-Kg at G.C.C.
Boosi Chettu	*Schleichera*	Bark	Rs 1/- Kg at G.C.C.
Chirimanu	*Terminalia paniculata*	Bark	Rs 1/- Kg at G.C.C.
Rella Chettu	*Cassia fistula*	Bark	Rs 2/- Kg G.C.C.
Yerra chekum	Pterolobium hexapetalum	Bark	1-2/Rs Kg at G.C.C.
Yelma or chimella chettu	*Feronia elephantum*	Bark	Rs 2/- Kg at G.C.C.
Tellamaddi	Arjunatree (*Terminalia arjuna*)	Bark	One rupee Kg in G.C.C.
Ganuga Chetti	*Pongamia pinnata*	Seed endosperm	Rupees 6/- Kg at G.C.C
Thene	Honey	Honey	Rs 80/-kg at G.C.C
Adda Chettu	*Bauhinia vahlili*	Leaves and fibre	Leaves Rs 2.50 per 100 and fiber Rs. 5/- Kg at G.C.C
Neradu chettu	Syzygium gumini var cununim	Fruits and wood	Fruits Rs. 1/-Kg at G.C.C. and tree wood is Rs. 200-300/-pillar to non-tribal
Mamidi Chettu	*Mangifera indica*	Fruit and wood	Parka Rs. 16/- broker, and wood is 250/- Rs. pair at private Businessman.

Contd...

Local name of the tree	*English Name*	*Parts collected*	*Sold@per Sack/Kg*
Panasa chettu	*Artocarpus integrefolicus*	Fruits and wood	One fruit Rs. 10-20/- private business man and large wood pillar is Rs. 300-400/-non-tribal
Jeedi mamidi	Cashew nut / *Anacardium occidentalis*	Fruit seeds	Rupees 30-35/- Kg sale to broker
Tuniki Chettu	Tobacco tree/*Diospyros melanoxylon*	Leaves	100 leaves per Rs 1.25 sale Private business man
Yelaga	*Limonia clephantum*	Fruits	100 fruits for Rs 200/- private business
Sitafalam	*Annona squamosa*	Fruit	Rupees. 50/- per 100 private business
Gummudu Chettu	*Cissis vistislinnae*	Wood for preparing Kadi	Fruit Rupees 60-100/- kg sale to broker and only wood for 400-500/- non-tribal
Tati Chettu	*Boraruss palm*	Leaves, fruits, wood, fiber of petiole, toddy, and sprouts.	Rs. 50/- per 100 Leaves, fiber Rs 8-9/-Kg, tree wood for Rs 300-500/-pillar, toddy Rs 10/-per litre sprouts Rs 20/-per 100 private businesses.
Jeelugu/jeeguru chettu	Caryota sap/ *Aeschynomene indica*	Sap/toddy	Toddy Rs 5/- half litre private business.
Yegesi	*Pterocarpus marsupium var bilous*	Wood preparing for plough,	For making plough Rs 40-60/- and wood for is Rs 250/- private business.
Veduru	Bamboo they prepare	Whole bamboo baskets and sale	Tadaka Rs. 5/- Tatlu Rs 10/- Gudu Rs. 50/- Buttalu Rs. 30-35/- and Gadde Rs. 400-500/- private business
Jaffrachettu	Bixa orellana	Seeds	Rs 90 to 100/ Kg at G.C.C.
Naditaram Chettu		Leaves	Rs 2.50 per 100 at G.C.C.
Chiguruaku		Leaves	Rs. 2.50 per 100 at G.C.C.

Appendix 18
Provides Information about Plant Species Used in Rituals and other Ceremonies by the Reddis

Local name	*Scientific name*	*Plant and it parts used for*
Vyapa chettu	*Azardirictha indica*	*Gangalamma* Goddess abode under this tree in the village *Utlapalem*
Burugu chettu	*Bombax cieba*	*Gangalamma* Goddess abode under this tree in the forest
Chinta chettu	*Tamarindis indica*	*Gangalamma* deity abode under this tree in the village Chintakarapalem
Velaga chettu	*Limonia crenulata*	*Gangalamma* deity abode under this tree in the village Nimmalapalem
Veduru pidam	Bamboo Grove/ *Bambusa arundiancea*	*Gangalamma* deity abode under this tree in the village Tirumalavada
Chinta chettu	*Tamarindis indica*	*Chintalamma* deity abode under this tree in the village Utalapalem
Raavi chettu	*Ficus riligiosa*	*Padalamma* deity abode under this tree in the village Raipalli
Rella chettu	*Cassia fistula*	Flowers of plant used for *Gangalamma* ritual
Pala chettu	*Mimusops hexandra*	Flowers of plant used for *Gangalamma* ritual and also leaves used for death rituals
Juvichettu or *Konda* juvvi *Tenkarai chettu*	*Ficus bengamina*	Bairudu abode God under this tree in the forest. Pothuraju God abode under this tree in the forest
Somi	*Soymida febrifuga*	Wood is used for making idols and temple pillars
Raavi chettu	*Urostigma religiosa*	Tree wood and leaves are use on the occasion of marriage.
Neredu chettu	*Syzygium cuminii*	The leaves and wood is used during marriage ceremony.
Mamdi chettu	*Mangifera indica*	Leaves of the tree used in during all festivals and rituals.
Pasapu	*Curcuma longa*	Fruit used in all rituals
Jeelugu/ jeerugu chettu	*Caryota urens*	Tree worshiped as chettu panduga
Tati chettu	*Borassus palm*	Tree worshiped as chettu panduga

Appendix 19
Provides Summary of Findings Key Characteristics of Indigenous Institutions

Category of indigenous institution	*Types of Institutions*	*Key Charateristics*
Social institutions	Kinship, clan, family, and land tenure systems	1. Network of relationships extend to the dead, unborn, and totems 2. emphasize collective decision making 3. communal ownership of land
Religious institutions	Ancestral worship, divinity and deities or gods	4. permeate all aspects of life 5. form the basis of morality 6. dictate code of conduct
Political institutions	Chieftaincy system at the family, community, village etc.	7. hierarchical levels of authority 8. custodians of lands 9. transactive decision making
Judicial institutions	Native court system at family, community village etd.	10. settle internal disputes and litigations 11. guided by precedents and wisdom of elders
Economic institutions	Indigenous healing, farming and hunting, wood carving, blacksmithing, basket weaving, and toddy tapping. These are indigenous occupation within indigenous institutions.	12. rooted in local culture and social values 13. experimental knowledge 14. have appropriate technologies for primary resource utilization 15. uphold holistic view of nature

Appendix 20
Provides Summary of Findings Indigenous Beliefs and Practices

Indigenous beliefs
The earth has a power of its own which is helpful if propitiated or appeased, and harmful if abused or neglected.
Ancestors are spirits who constantly observe the behaviour of the living.
Ancestors reward those who act in accordance with societal values and punish those who exhibit deviant behaviours.
Spirits dwell in such natural resources as hills, forests, rocks, water bodies and in certain animals.
Every creature is endowed with a soul, which survives after death.
Certain animals and plant species have vindictive souls.
There is great respect for the earth farming activities are suspended on sacred days.
Elders are respected and their age and experimental knowledge are recognized.
In case of game, *Reddis* do not eat the meat of the animal, which is attacked by the tiger
Meats of Kanasu, Pandi (wild boar), Chevulapilli (rabbita), are not eaten by inhabitants of villagers who are suffering from any diseases.
It is taboo for women to step into the fields or near the God and Goddess during the menstrual period.

Bibliography

Abdel Ghaffar Ahmed, M. (1994), The Relevance of Indigenous Systems of Production to Food Policies: A Horn of Africa Perspective. *Eastern Africa Social Science Research Review (2):54-65.*

Action Plan (2000), Konda Reddi Action Plan. ITDA Rampachoudavaram. Agrawal, A. 1995. Dismantling the divide between Indigenous and Scientific Knowledge. *Development and Change* 26.

Agrawal, Arun (1993), "Mobility and Control among Nomadic Shepherds: The Case of the Raikas". Human Ecology, 21: 261-79.

Anderson, D. and Grove, R. (1987), "Conservation in Africa" People, Policies and Practice, Cambridge, Cambridge University Press.

Ann Fienup-Riordan (1990), Original Ecologists? The Relationship between Yup'ik Eskimos and Animals, in *Eskimo Essays*, A. Fienup-Riordan, ed. London: Rutgers UP

Antweiler, christoph (2004), Local Knowledge Theory and Methods: An Urban Model from Indonesia. In *Investigating Local Knowledge: New Directions, New Approaches.* (ed.) Alan Bicker, Paul Sillitoe, Johan Pottier. England: Ashgate Publishing Limited.

Atran, Scott (1991), "Ethnoscience Today." Social Science Information 30: 595-62.

Attfield, Robin, and Andrew Belsey (1994), Introduction to *Philosophy and the natural environment*, edited by Robin Attfield and Andrew Belsey, 1-13. Cambridge: Cambridge University Press.

Ayers, Alison (1995), Indigenous soil and water conservation in Djenne, Mali. In *The cultural dimension of development: indigenous knowledge systems*, by D. Michael Warren *et al.*, 371-384. London: Intermediate Technology Publications.

Babiker, Mustafa (1996), Management of Aridity: Water Conservation and Procurement in Dar Hamar, Western Sudan. In *Managing scarcity: Human adaptation in East African dry lands*, edited by Abdel Ghaffar M. Ahmed and Hassan A. Abdel Ati, 107-126. Proceedings of a Regional Workshop held on 24-26 August 1995. Addis Ababa: OSSREA.

Balick, M. and P. Cox (1997), Plants, People and Culture: The Science of Ethnobotany. Scientific American, NY.

Bandyopadhyay, A. K. and Saha, G. S. (1998), Indigenous Methods of Seed Selection and Preservation on the Andaman Islands in India. Indigenous Knowledge and Development Monitor. Vol. 6, 1:3-6.

Barrett, S. (1996), *Anthropology, A Students Guide to Theory and Method*. University of Toronto Press. Toronto, Buffalo, London.

Bartels, L. (1990), *Oromo religion: Myths and Rites of the Western Oromo of Ethiopia-An Attempt to Understand*. Berlin: Dietrich Reinner.

Bennet, Olivia, and Nigel Cross, eds. (1993), *Listening for a Change: Oral Testimony and Development*. London: Panos Publications.

Bennett, B. (1992), Plants and People of the Amazonian Rainforests: The Role of Ethnobotany in Sustainable Development. BioScience 42: 599-607.

Berks, Fikret *et al.* (1995), Traditional Ecological Knowledge, Biodiversity, Resilience and Sustainability. In *Biodiversity Conservation: Problems and Policies*, edited by C.A. Perrings *et al.* Dordrecht, Boston, London: Kluwer Academic Publishers.

Berlin, Brent (1992), Ethnobiological Classification: Principles of Categorization of Plants and Animals in Traditional Societies.

Brokensha, D., Warren, D and Werner, O. (1980), *Indigenous Knowledge Systems and Development*, University Press of America, Lanham.

Bronowski, J. (1978), *The Origins of Knowledge and Imagination*. New Haven and London, Yale University Press.

Brouwer, Jan (1998), Indigenous Knowledge, 'Indigenous Knowledge Systems', 'Indigenous Technological Knowledge'. *Indigenous Knowledge and Development Monitor*, Vol. 6., 3 p. 13.

Brush, S. (1980), "Potato Taxonomies in Andean Agriculture." Pp. 37-47 in D. Brokensha, D. Warren, and O. Werner (eds.),

Indigenous Knowledge Systems and Development. Lanham, MD: University Press of America.

Center for Indigenous Knowledge for Agriculture and Rural Development (CIKARD), established in 1987 at Iowa State University, Ames, Iowa USA.

Brush, S. (1993), Indigenous Knowledge of Biological Resources and Intellectual Property Rights: The Role of Anthropology. *Annual Reviews of Anthropology,* 95(3).

Chadwick, D. and J. Marsh (eds) (1994), Ethnobotany and the Search for New Drugs. Wiley, Chichister.

Chambers, Robert (1979) (ed.) *Rural Development: Whose Knowledge Counts?* IDS Bulleton, 10(2)

Chambers, Robert (1983), *Rural Development: Putting the Last First.* Longman, London.

Chandrakanth, M. G. and Romm, Jeff (1991), Sacred Forests, Secular Forest Policies and People's Actions. *Natural Resources Journal,* 31, 741-756.

Clarke, W. C. (1977), The Structure of Permanence: The Relevance of Self-subsistence Communities for World Ecosystem Management. In *Subsistence and Survival: Rural Ecology in the Pacific,* edited by T. Bayliss-Smith and R. Feachem, 363-84. London: Academic Press.

Cohen, John M. (1987), *Integrated Rural Development: The Ethiopian Experience and the Debate.* Uppsala: The Scandinavian Institute of African Studies.

Conklin, H. C. (1957), Hanunoo Agriculture. A Report on an Integral System of Shifting Cultivation in the Philippines. Forestry Development Paper No. 5. FAO Rome.

Cunningham, A. (1996), Professional Ethics and Ethnobotanical Research. *In* Alexiades and Sheldon.

Cunningham, A. (2001), Applied Ethnobotany. Earthscan, London.

Das, Harish Chandra (1997), Man's Relationship with Forest: Deification of Trees and Plants, In: Seeland, Klaus and Schmithusen, Franz (eds.) *Local Knowledge of Forests and Forest Uses among Tribal Communities in In India,* Department Wald-und Holzforschung, Zurich.

Dasmann, R. F. (1991), The Importance of Cultural and Biological Diversity. In Biodiversity: Culture, Conservation, and

Ecodevelopment, Margery L. Oldfield and Janis B. Alcorn, Eds. 7-15. Boulder: Westview Press.

Davis, Shelton (ed). (1993), World Bank Discussion Paper 188: Indigenous Views of the Land and the Environment. The World Bank: Washington D.C.

Dei, George Jerry Sefa, Budd L. Hall and Dorothy Goldin Rosenberg. (2000), Indigenous Knowledges in Global Contexts: Multiple Readings of Our World. OISE/UT Book/ University of Toronto Press. Toronto.

Dessalegn Rahmato (1997), Environmentalism and Conservation in Wallo before the Revolution. Paper Presented at the Conference on Environment and Development in Ethiopia, Debre Zeit, 12-15 June.

DeWalt, B. R. (1994), Using Indigenous Knowledge to Improve Agriculture and Natural Resource Management. *Human Organization*, 53: (2): 123-131.

Eghenter, C. (2000), What is Tana Ulen good for? Considerations on Indigenous Forest Management, Conservation and Research in the Interior of Indonesian Borneo. *Human Ecology*, 28, 3, 331.

Ellen, R. F. and Holly Harris (2000), Introduction, In Ellen, R. F., Peter Parkes and Alan Bicker (eds). 2000. *Indigenous Ecological Knowledge and its Transformations*. Amsterdam: Harwood Academic.

Ellen, Roy (1998), "Doubts About a Unified Cognitive Theory of Taxonomic Knowledge and its Memic Status." Behavioral and Brain Sciences, 21, 572.

Ellen, Roy F. (1978), *Nuaulu Settlement and Ecology: An Approach to the Environmental Relations of an Eastern Indonesian Community*. The Hague: Martinus Nijhoff. xi+265.

Ellen, Roy (2002), 'Déjà vu, All Over Again', Again: Reinvention and Progress in Applying Local Knowledge to Development. In *Participating in Development: Approaches to indigenous knowledge*. (ed.) Paul Sillitoe, Alan Bicker and Johan Pottier. London: Routledge.

Farnsworth, N. and D. Soejarto (1985), Potential Consequence of Plant Estinction in the US on the Current and Future Availability of Prescription Drugs. Econ. Bot., 39: 231-240.

Fathy, H. (1986), *Natural Energy and Vernacular Architecture*. Chicago, University of Chicago Press.

Feldstein, H. and Jiggins, J. (1994), *Tools for the Field: Methodologies Handbook for Gender Analysis in Agriculture*. Kumarian Press, West Hartford.

Fernandez-Baca, Edith (1994), The Role of Women in Livestock Production in the Mantaro Valley (Peru). Indigenous Knowledge Monitor. Special Issue 2(3) *http://www.nuffic.nl/ciran/ikdm/2-3/contents.html*

FIKENTSCHER, Wolfgang (1995), Modes of Thought. A Study in the Anthropology of Law and Religion. Tübingen: J.C.B. Mohr (Paul Siebeck).

Flora, Cornelia B. (1992), "Reconstructing Agriculture: The Case for Local Knowledge". *Rural Sociology*, 57 1, 92-97.

Freeman, M. M. R., Y. Matsuda, and K. Ruddle, eds. (1991), *Adaptive Marine Resource Management Systems in the Pacific*. Resource Management and Optimization 8. No.3/4.

Furer-Haimendrof, C von & Elizabeth Furer-Haimendrof (1945), *The Reddis of Bison Hills: A Study in Acculturation*, London: Macmillan & Co. Ltd.

Gadgil and Guha (1992), State Forestry and Social Conflict in British India. In David Hardiman (ed) Peasant Resistance in India, 1858-1914, Delhi: Oxford University Press.

Gadgil, M. and F. Berkes (1991), *Traditional Resource Management Systems*. Resource Management and Optimization, 18: 127-141.

Gadgil, Madhav and Berkes, F. and Folke, Carl (1993), Indigenous Knowledge for Biodiversity Conservation. *Ambio*. 22, (2-3), 266-270.

Gadgil, Madhav and Guha, Ramachandra (1992), *This Fissured Land: An Ecological History of India*. OUP, New Delhi.

Gadgil, Madhav and Vartak, V. D. (1976a), Sacred Groves in India: A Plea for Continuous Conservation. *J. Bombay nat. Hist. Soc.*, 72, 314-320.

Gegeo, D. (1998), Indigenous Knowledge and Empowerment: Rural Development Examined from Within. *The Contemporary Pacific*, 10, (2), 289-316.

Grenier, L. (1998), Working with Indigenous Knowledge: A Guide for Researches. Ottawa: IDRC.

Guha, R. (1994), "Social Ecology" Oxford in India Readings in Sociology and Social Anthropology. New Delhi, Oxford University Press.

Hardesty, D. L. (1977), *Ecological Anthropology*. New York, Wiley.

Howes, M. (1980), The Uses of Indigenous Technical Knowledge in Development. In *Indigenous Knowledge Systems and Development*, Edited by D. Warren Brokensha, Wernerd D. Lanham: University Press of America, 335-351.

Hultin, J. (1987), *The Long Journey: Essays on History, Descent and Land among the Macha Oromo*. Uppsala: Uppsala University.

Hultin, J. (1994), The Land is Crying: State Intervention and Cultural Resistance. In *A River of Blessings: Essays in Honour of Paul Baxter*, edited by D. Brokensha, 67-83. New York: Maxwell School of Citizenship and Public Affairs.

IDS (1979), Rural Development: Whose Knowledge Counts? *IDS Bulletin*, 10 (2).

IUCN (International Union for the Conservation of Nature and Natural Resources), 1991. Caring for the Earth: A Strategy for Sustainable Living. IUCN, United Nations Environment Programme, and World Wide Fund for Nature, Gland, Switzerland.

Iwanaga, Masa (1998), "*In situ* conservation and the development process" Pp.vi in Jarvis, Debra I. and Toby Hodgkin (Eds.). Strengthening the scientific basis of *in situ* conservation of agricultural biodiversity on-farm: options for data collecting and analysis. Proceedings of a workshop to develop tools and procedures for *in situ* conservation on- farm, 25- 29 august 1997, Rome, Italy. International Plant Genetic Resources Institute, Rome. Italy.

Jain, S. K. and Sneh Lata (1996), Unique Indigenous Amazonian Uses of some Plants Growing in India. *Indigenous Knowledge and Development Monitor*, Vol. 4, 3, 21-23.

Jena, Mihir Kumar (1997), Forest and Sacred Tribal Complex: A Comparison of Three Tribes of Orissa, In: Seeland, Klaus and Schmithusen, Franz (eds.) *Local Knowledge of Forests and Forest Uses among Tribal Communities in India*, Department Wald-und Holzforschung, Zurich..

Jodha, N. S. (1986), Common Property Resources and Rural Poor in Dry Regions of India. *Economic and Political Weekly*, Vol. XXI (27): 1169-1181.

Johannes, R.E., ed. (1989), *Traditional Ecological Knowledge: A Collection of Essays*. Gland, Switzerland and Cambridge, UK, IUCN.

Johnson, Martha (ed). (1992), *LORE: Capturing Traditional Environmental Knowledge*. Ottawa (CA): Dene Cultural Institute & International Development Research Centre.

Kassam, Aneesa, and Gamatchu Megersa (1994), Aloof Alollaa: The Inside and the Outside-Boran Oromo Environmental Law and Methods of Conservation. In *A River of Blessings: Essays in Honour of Paul Baxter*, edited by D. Brokensha, 85-98. New York: Maxwell School of Citizenship and Public Affairs.

Kebede Tato and Hans Hurni (1992), Introduction to *Soil Conservation for Survival*, edited by Kebede Tato and Hans Hurni, 1-10. Ankey: Soil and Water Conservation Society.

Keith Thomas (1983), *Man and theNatural World*. Oxford UP.

Klemm, C. de. (1985), Culture and Conservation: Some Thoughts for the Future. In *Culture and Conservation: The Human Dimension in Environmental Planning*, edited by Jeffrey A. McNeely and David Pitt, 239-257. London/New York: Croom Helm.

Kloppenburg, Jack (1991), "Social Theory and the De/reconstruction of Agricultural Science: Local Knowledge for an Alternative Agriculture." *Rural Sociology*, 56, 4, 519-548.

KNIGHT, C. Gregory (1980), Ethnoscience and the African Farmer: Rationale and Strategy. In: BROKENSHA, David *et al!* (eds.):203-229.

Knudtson, P., and D. Suzuki (1992), *Wisdom of the Elders*. Toronto, Stoddart.

Knutsson, K. E. (1967), *Authority and Change: A Study of the Kallu Institution among the Macha Galla of Ethiopia*. Gothenburg: Ethnografiska Museet.

Kothari, A., Neema Pathak, Anuradha R. V., Bansuri Taneja (1998), Communities and Conservation: Natural Resource Management in South and Central Asia. New Delhi, Sage Publication.

Madhav Gadgil (1991), Conserving India's Biodiversity: The Societal Context. Evolutionary Trends in Plants 5 (1): 3-8.

Maikhuri, R. K., Sunil Nautiyal, K. S. Rao and R. L. Semwal (2000), Indigenous Knowledge of Medicinal Plants and wild edibles among three Tribal Sub-communities of the Central Himalayas, India. *Indigenous Knowledge and Development Monitor*. Vol. 8, 2, 3-7-13.

Mann, R. S. (1990), Culture and Ecology in Ladakha. Man in India. Vol. 70 (3) September.

Maurial, Mahia (1999), "Indigenous Knowledge and Schooling: A Continuum between Conflict and Dialogue" Pp. 59-77 in Semali, Ladislaus M. and Joe L. Kincheloe (Eds.). What is indigenous knowledge? Voices from the academy. New York and London. Falmer Press.

Mbaiwa, J. E. (2002), The Socio-Economic and Environmental Impact of Tourism Development in the Okavango Delta, Botswana: A Baseline study, Harry Oppenheimer Okavango Research Centre, University of Botswana, Maun.

Mc Corkle, Constance M. (1989), Toward a Knowledge of Local Knowledge and its Importance for Agricultural RD&E. *Agriculture and Human Values*, Summer, 4-12.

McNeely, Jeffrey A. (1995), "IUCN and Indigenous Peoples: How to Promote Sustainable Development", p. 445- 450 in Warren, D. Michael, L. Jan Slikkerveer and David Brokensha (Eds.). The Cultural Dimension of Development: Indigenous Knowledge Systems. London: Intermediate Technology Publications. UK.

Meehan, P. (1980), 'Science, Ethoscience and Agricultural Knowledge Utilization; in D. Brokensha, D. Warren and O.Werner (eds), Indigenous Knowledge Systems and Development, 383-92 Lanham University of Chicago Press.

Misra, Kamal K. & K. K. Basa (2005), Indigenous Knowledge, Natural Resource Management and Development. The Konda Reddi Experience, IGRMS Bhopal & Pratibha Prakashan, New Delhi.

Misra, Kamal K. (1998), Trees in the Religion and Folklore of the Konda Reddi of South India. *Journal of Human Ecology*, 9 (3).

Mohamed, Yagoub Abdella (1994), Biodiversity Considerations among Traditional Farmers of Western Sudan. In *Indigenous Farming Systems, Knowledge and Practices in the Sudan*, edited by Medani Mohamed M Ahmed, 167-190. Khartoum: Khartoum University Press.

Murali, V., Venkaiah, K, Satyanarayana, V.V.V. (2000), Palms of Eastern Ghats and their Utilization. *EPTRI-ENVIS Newsletter*, Vol. (6): 1:6-7.

MURDOCH, Jonathan and CLARK, Judy (1994), Sustainable Knowledge. *Geoforum*, 25, 2, 115-132.

OLSON, David R. and TORRANCE, Nancy (eds.) (1996), Modes of Thought. Explorations in Culture and Cognition. Cambridge etc.: Cambridge University Press.

Omari, C. K. (1990), Traditional African Land Ethics. In *Ethics of Environment and Development: Global Challenge, International Response*, edited by J. R. Engel and Joan Gibb Engel, 167-75. London: Belhaven Press.

Orlove, S. Benjamin (1980). "Ecological Anthropology". *Annual Review of Anthropology,* Vol 5, 235-63.

Ostrom, E. (1990), Governing the Commons: The Evolution of Institutions for Collective Action, Cambridge University Press, Cambridge.

Pandey, Deep N. (1994b), *Wild Trees of Aravallis.* Himanshu Publications, Udaipur/Delhi.

Pandey, Deep N. (1996a), Participatory Forest Management in Udaipur. *Yojana,* 40(7): 18-20

PELISSIER, Catherine (1991), The Anthropology of Teaching and Learning. *Annual Review of Anthropology,* 20, 75-95

Posey, D. (2004), *Indigenous Knowledge and Ethics: A Darrell Posey Reader.* London: Routledge.

Posey, D.A. (1985), *Indigenous Management of Tropical Forest Ecosystems: The Case of the Kayapo Indians of the Brazilian Amazon.* Agroforestry Systems 3:139-158.

Posey, D.A. (1991), 'Kayap' Indians: Experts in Synergy, *ILEIA Newsletter,* 7(4):3-5.

Posey, Darrell (2002), Upsetting the Sacred Balance: Can the Study of Indigenous Knowledge Reflect Cosmic Connectedness? In *Participating in Development: Approaches to indigenous knowledge.* (ed.) Paul Sillitoe, Alan Bicker and Johan Pottier. London: Routledge.

RAFI (2000), In search of higher ground: the intellectual property challenge to public agricultural research and human rights and 28 alternative initiatives. *The Occasional Paper Series,* Vol. 6, No. 1.

RAFI/ UNDP (1995), CSOPP Documents: Conserving Indigenous Knowledge-Integrating New Systems of Integration. Web Site. *http://www.undp.org/csopp/*

Ramakrishanan, P. S. (1992), Jhum: Is There a Way out? In B. Chaudhuri (ed) Tribal Transformation in India, New Delhi: Inter India Publication.

Rao, J. V. Ramana (2000), Non Timber Products (Minor Forest Produce) Utilization by Different Tribals at Mahadevpur in Karimnagar District in the context of Ecological Change and Conservation of Biodiversity, Final Technical Report Submitted to Ministry of Environment and Forests, Government of India.

Reed C. (1977), Origins of Agriculture. The Hague: Mouton.

Rhoades, R. (1989), "The Role of Farmers in the Creation of Appropriate Technology." Pp. 3-9 in R. Chambers, R. Pacey, and L. Thrupp (eds.), Farmer First: Farmer Innovation and Agricultural Research. London: Intermediate Technology Publications.

Richards, P., ed. (1975), *African Environment, Problems and Perspectives.* London: International African Institute.

Roose, Eric. (1992), Traditional and Modern Strategies for Soil and Water Conservation in the Sudano-Sahelian Areas of Western Africa. In *Erosion, Conservation, and Small-scale Farming,* edited by Hans Hurni and Kebede Tato, 349-365. Missouri: Wadsworth Publishing Company.

Roy Burman, J. J. (1995), 'The Dynamics of Sacred Groves; *Journal of Human Ecology,* Vol. 6, 245-54.

Salau, Ademola T. (1992), *Global Environmental Change: A Research Agenda for Africa.* Dakar: CODESRIA. Working Paper No. 2.

Salih, M. M. M. (1992), *Pastoralists and Planners: Local Knowledge and Resource Management in Gidan Magajia Grazing Reserve, Northern Nigeria.* London: IIED.

Schmithusen, Franz (1997), Local Knowledge on Forests, In: Seeland, Klaus and Schmithusen, Franz (eds.) *Local Knowledge of Forests and Forest Uses among Tribal Communities in India,* Department Wald-und Holzforschung, Zurich.

Seeland, Klaus (1997), Nature is Culture: Indigenous Knowledge and Soico-cultural Aspects of Trees and Forests in Non-Eurpean Cultures. London; Intermediate Technology Publications.

Seeland, Klaus (1997), What is Indigenous Knowledge and Why Does it Matter Today? In : Seeland, Klaus and Schmithusen, Franz (eds.) Local Knowledge of Forests and Forest Uses among Tribal Communities in India, Department Wald-und Holzforschung, Zurich.

Semali, Ladislaus M. and Joe L. Kincheloe (Eds.). (1999), What is indigenous knowledge? Voices from the academy. New York and London. Falmer press.

Sillitoe, P. (1998), The Development of Indigenous Knowledge: A New Applied Anthropology. *Current Anthropology*, 39, 2, 223-252.

Sillitoe, Paul and Alan Bicker (2004), Introduction: Hunting for Theory, Gathering Ideology. In *Development and Local Knowledge*, (ed.) Alan Bicker, Paul Sillitoe and Johan Pottier. London: Routledge.

Sillitoe, Paul (2002b), Globalizing Indigenous Knowledge. In *Participating in Development: Approaches to Indigenous Knowledge*. (ed.) Paul Sillitoe, Alan Bicker and Johan Pottier. London: Routledge.

Sillitoe, Paul (2002a), Participant Observation to Participatory Development: Making Anthropology Work. In *Participating in Development: Approaches to indigenous knowledge*. (ed.) Paul Sillitoe, Alan Bicker and Johan Pottier. London: Routledge.

Singh, Katar (1994), Managing Common Pool Resources: Principles and Case Studies, Oxford UP, New Delhi.

Singh, V. and Pandey, R. P. (1982), Plants Used in Religion and Magico-religious Beliefs in Rajasthan. *J.Ecor. Tax. Bot.*, 3, 273-278.

Slocum, R. *et al.* (1995), *Power, Process and Participation: Tools for Change*. Intermediate Technology Publications, London.

Sunitha Reddy (2004), Ecosystems Approach to Human Health: A Case of Konda Reddi Tribes and Women's Health. *Journal of Human Ecology*, 16 (4): 271-282.

Steward, J. H. (1955), Theory of Culture Change. Urbana, University of Illinois Press.

ten Kate, K. and S. Laird (1999), The Commercial Use of Biodiversity: Access to Genetic Resources and Benefit-sharing. Earthscan, London.

Thrupp, L. A. (1989), Legitimizing Local Knowledge, 'Scientized Packages' or Empowerment for Third World People. In *Indigenous Knowledge Systems: Implications for Agriculture and International Development*, edited by D. M. Warren, *et al.* Iowa: Iowa State University.

Vandana Shiva (1991), Ecology and the Politics of Survival, Conflicts over Natural Resources in India. United Nations University Press, New Delhi.

Vatsyayan, Kapila (1992), Ecology and Indian Myth. In, Geeti Sen(ed.) *IIndigenous Vision*. Sage and IIC, New Delhi.

Viergever, Marcel (1999), "Indigenous Knowledge: An Interpretation of Views from Indigenous Peoples". Pp. 333-359 in Semali, Ladislaus M. and Joe L. Kincheloe (Eds.). What is Indigenous Knowledge? Voices from the Academy. New York and London. Falmer Press.

Wade, R. (1986), Common Property Resource Management in South India Village, in Proceeding of the Conference on Common Property Resource Management, April 21-26, Board of Science and Technology for International Development Affairs, National Research Council, Washington D.C.

Warren, D. M., Slikkerveer, L. J. and Brokensha, D. (1995), *The Cultural Dimensions of Development, Indigenous Knowledge Systems*. London: Intermediate Technology Publication.

Warren, D. Michael and Gerard Mckiernan (1995), 'CIKARD: A Global Approach to Documenting Indigenous Knowledge for Development Pp. 426-434 in Warren, D. Michael, L. Jan Slikkerveer and David Brokensha (Eds.). The Cultural Dimension of Development: Indigenous Knowledge Systems. London: Intermediate Technology Publications. UK.

D.M. Warren (1992), **Preliminary Analysis of Indigenous Soil Classification and Management Systems in Four Ecozones of Nigeria**. Nigeria, ARCIK/NISER, in collaboration with the International Institute of Tropical Agriculture (IITA) 28 pp.

WARREN, D. M., SLIKKERVEER, L. J. and TITILOLA, S. O. (eds) (1989), Indigenous Knowledge Systems: Implications for Agriculture and International Development. *Studies in Technology and Social Change Program*, No. 11. Ames, Iowa: Iowa State University Research Foundation.

WCED (1987), *Our Common Future*. The World Commission on Environment and Development. Oxford and New York, Oxford University Press.

Zent, Stanford (2000), "A Genealogy of Scientific Perspectives of Indigenous Knowledge." Paper prepared for White Oak Seminar.

Photographs

Ploughing in Progress

Preparation of Brinjal for Crop Plantation

Sun-draying of Faram Dumpa Tuber (Crop)

Collection of Bamboo from Forest for House Construction

Fibre Preparation of Palm Petiole (Tati Balage)

Cutting of Palm Leaves for Roof Construction

Indigenous Hunting Practice Ceremony

Villager Carrying Wild Animal (Kanasu) after Hunting

Trading Village Council Meeting at Utlapalem

Gangalamma, the Village Deity Place of Worship and Offering

Kola Dance Celebrating on the Occasion of Gangalamma Festival

Celebrating of Gangalamma Festival

Maps

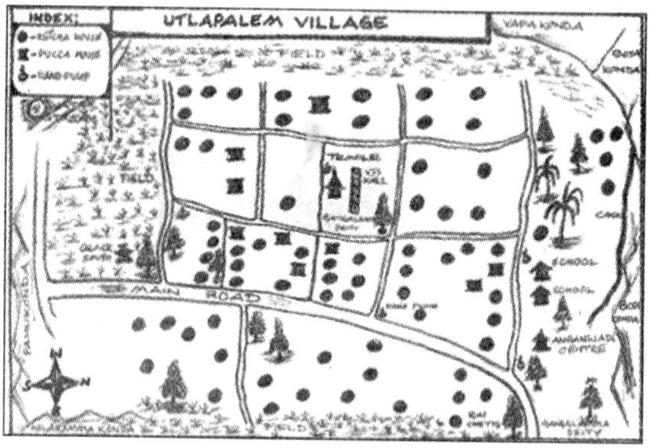

Map of State Andhra Pradesh, District East Godavari and Village Utlapalem

Index